NARRATIVE OF THE MUTINY BOUNTY

WILLIAM BLIGH

ISBN: 151516974X
ISBN-13: 978-1515169741

NARRATIVE OF THE MUTINY BOUNTY

On Board His Majesty's Ship Bounty; and the Subsequent Voyage of Part of the Crew, in the Ship's Boat, From Tofoa, one of the Friendly Islands, To Timor, a Dutch Settlement in the East Indies

LIST OF ILLUSTRATIONS

ADVERTISEMENT.

The following Narrative is only a part of a voyage undertaken for the purpose of conveying the Bread-fruit Tree from the South Sea Islands to the West Indies. The manner in which this expedition miscarried, with the subsequent transactions and events, are here related. This part of the voyage is not first in the order of time, yet the circumstances are so distinct from that by which it was preceded, that it appears unnecessary to delay giving as much early information as possible concerning so extraordinary an event. The rest will be laid before the Public as soon as it can be got ready; and it is intended to publish it in such a manner, as, with the present Narrative, will make the account of the voyage compleat.

At present, for the better understanding the following pages, it is sufficient to inform the reader, that in August, 1787, I was appointed to command the Bounty, a ship of 215 tons burthen, carrying 4 six-pounders, 4 swivels, and 46 men, including myself and every person on board. We sailed from England in December, 1787, and arrived at Otaheite the 26th of October, 1788. On the 4th of April, 1789, we left Otaheite, with every favourable appearance of completing the object of the voyage, in a manner equal to my most sanguine expectations. At this period the ensuing Narrative commences.

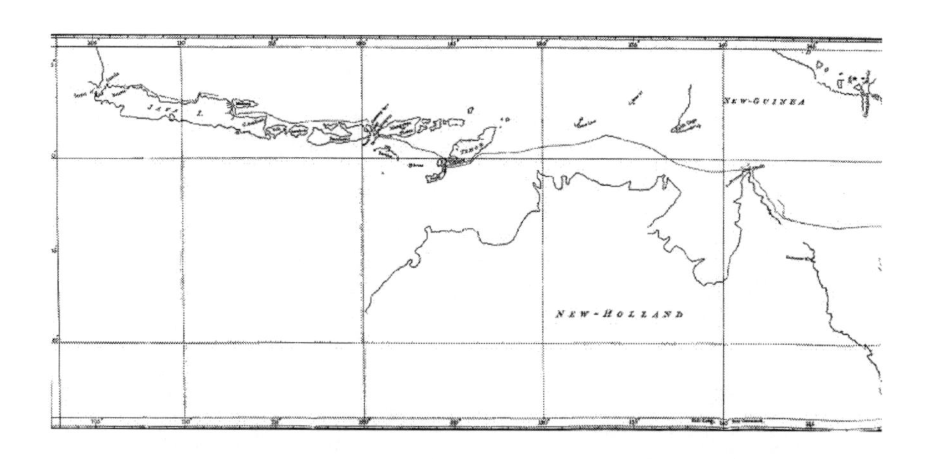

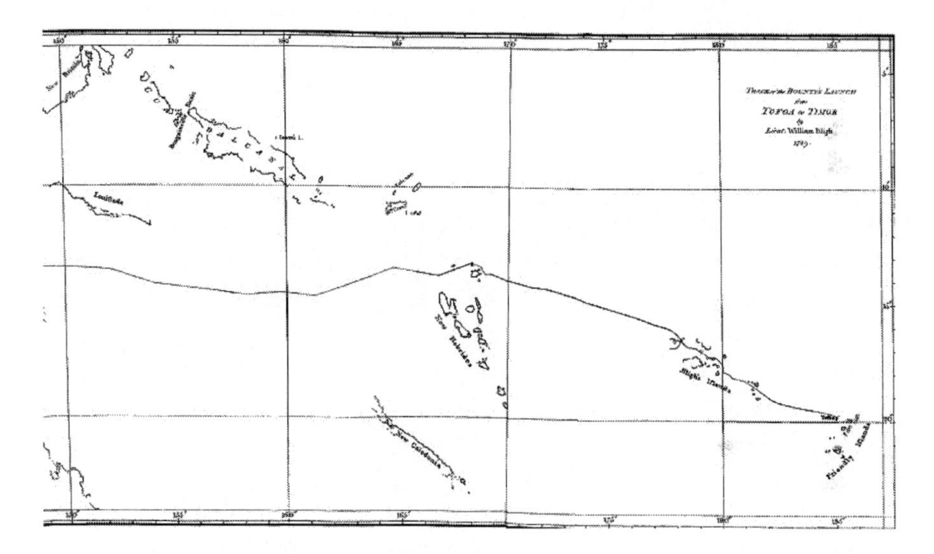

Track of the Bounty's Launch from Tofoa to Timor by Lieut. William Bligh, 1789

A

NARRATIVE, &c.

APRIL

1789. April.

I sailed from Otaheite on the 4th of April 1789, having on board 1015 fine bread-fruit plants, besides many other valuable fruits of that country, which, with unremitting attention, we had been collecting for three and twenty weeks, and which were now in the highest state of perfection.

On the 11th of April, I discovered an island in latitude 18° 52´ S. and longitude 200° 19´ E. by the natives called Whytootackee. On the 24th we anchored at Annamooka, one of the Friendly Islands; from which, after completing our wood and water, I sailed on the 27th, having every reason to expect, from the fine condition of the plants, that they would continue healthy.

On the evening of the 28th, owing to light winds, we were not clear of the islands, and at night I directed my course towards Tofoa. The master had the first watch; the gunner the middle watch; and Mr. Christian, one of the mates, the morning watch. This was the turn of duty for the night.

1789. April.

Just before sun-rising, Mr. Christian, with the master at arms, gunner's mate, and Thomas Burket, seaman, came into my cabin while I was asleep, and seizing me, tied my hands with a cord behind my back, and threatened me with instant death, if I spoke or made the least noise: I, however, called so loud as to alarm every one; but they had already secured the officers who were not of their party, by placing centinels at their doors. There were three men at my cabin door, besides the four within; Christian had only a cutlass in his hand, the others had muskets and bayonets. I was hauled out of bed, and forced on deck in my shirt, suffering great pain from the tightness with which they had tied my hands. I demanded the reason of such violence, but received no other answer than threats of instant death, if I did not hold my tongue. Mr. Elphinston, the master's mate, was kept in his birth; Mr. Nelson, botanist, Mr. Peckover, gunner, Mr. Ledward, surgeon, and the master, were confined to their cabins; and also the clerk, Mr. Samuel, but he soon obtained leave to come on deck. The fore hatchway was guarded by centinels; the boatswain and carpenter were, however, allowed to come on deck, where they saw me standing abaft the mizen-mast, with my hands tied behind my back, under a guard, with Christian at their head.

The boatswain was now ordered to hoist the launch out, with a threat, if he did not do it instantly, to take care of himself.

The boat being out, Mr. Hayward and Mr. Hallet, midshipmen, and Mr. Samuel, were ordered into it; upon which I demanded the cause of such an order, and endeavoured to persuade some one to a sense of duty; but it was to no effect: "Hold your tongue, Sir, or you are dead this instant," was constantly repeated to me.

The master, by this time, had sent to be allowed to come on deck, which was permitted; but he was soon ordered back again to his cabin.

1789. April.

I continued my endeavours to turn the tide of affairs, when Christian changed the cutlass he had in his hand for a bayonet, that was brought to him, and, holding me with a strong gripe by the cord that tied my hands, he with many oaths threatened to kill me immediately if I would not be quiet: the villains round me had their pieces cocked and bayonets fixed. Particular people were now called on to go into the boat, and were hurried over the side: whence I concluded that with these people I was to be set adrift.

I therefore made another effort to bring about a change, but with no other effect than to be threatened with having my brains blown out.

The boatswain and seamen, who were to go in the boat, were allowed to collect twine, canvas, lines, sails, cordage, an eight and twenty gallon cask of water, and the carpenter to take his tool chest. Mr. Samuel got 150lbs of bread, with a small quantity of rum and wine. He also got a quadrant and compass into the boat; but was forbidden, on pain of death, to touch either map, ephemeris, book of astronomical observations, sextant, time-keeper, or any of my surveys or drawings.

The mutineers now hurried those they meant to get rid of into the boat. When most of them were in, Christian directed a dram to be served to each of his own crew. I now unhappily saw that nothing could be done to effect the recovery of the ship: there was no one to assist me, and every endeavour on my part was answered with threats of death.

1789. April.

The officers were called, and forced over the side into the boat, while I was kept apart from every one, abaft the mizen-mast; Christian, armed with a bayonet, holding me by the bandage that secured my hands. The guard round me had their pieces cocked, but, on my daring the ungrateful wretches to fire, they uncocked them.

Isaac Martin, one of the guard over me, I saw, had an inclination to assist me, and, as he fed me with shaddock, (my lips being quite parched with my endeavours to bring about a change) we explained our wishes to each other by our looks; but this being observed, Martin was instantly removed from me; his inclination then was to leave the ship, for which purpose he got into the boat; but with many threats they obliged him to return.

The armourer, Joseph Coleman, and the two carpenters, M'Intosh and Norman, were also kept contrary to their inclination; and they begged of me, after I was astern in the boat, to remember that they declared they had no hand in the transaction. Michael Byrne, I am told, likewise wanted to leave the ship.

It is of no moment for me to recount my endeavours to bring back the offenders to a sense of their duty: all I could do was by speaking to them in general; but my endeavours were of no avail, for I was kept securely bound, and no one but the guard suffered to come near me.

To Mr. Samuel I am indebted for securing my journals and commission, with some material ship papers. Without these I had nothing to certify what I had done, and my honour and character might have been suspected, without my possessing a proper document to have defended them. All this he did with great resolution, though guarded and strictly watched. He attempted to save the time-keeper, and a box with all my surveys, drawings, and remarks for fifteen years past, which were numerous; when he was hurried away, with "Damn your eyes, you are well off to get what you have."

1789. April.

It appeared to me, that Christian was some time in doubt whether he should keep the carpenter, or his mates; at length he determined on the latter, and the carpenter was ordered into the boat. He was permitted, but not without some opposition, to take his tool chest.

Much altercation took place among the mutinous crew during the whole business: some swore "I'll be damned if he does not find his way home, if he gets any thing with him," (meaning me); others, when the carpenter's chest was carrying away, "Damn my eyes, he will have a vessel built in a month." While others laughed at the helpless situation of the boat, being very deep, and so little room for those who were in her. As for Christian, he seemed meditating instant destruction on himself and every one.

I asked for arms, but they laughed at me, and said I was well acquainted with the people where I was going, and therefore did not want them; four cutlasses, however, were thrown into the boat, after we were veered astern.

1789. April.

When the officers and men, with whom I was suffered to have no communication, were put into the boat, they only waited for me, and the master at arms informed Christian of it; who then said—"Come, captain Bligh, your officers and men are now in the boat, and you must go with them; if you attempt to make the least resistance you will instantly be put to death:" and, without any farther ceremony, holding me by the cord that tied my hands, with a tribe of armed ruffians about me, I was forced over the side, where they untied my hands. Being in the boat we were veered astern by a rope. A few pieces of pork were then thrown to us, and some cloaths, also the cutlasses I have already mentioned; and it was now that the armourer and carpenters called out to me to remember that they had no hand in the transaction. After having undergone a great deal of ridicule, and been kept some time to make sport for these unfeeling wretches, we were at length cast adrift in the open ocean.

I had with me in the boat the following persons:

Names.	**Stations.**
John Fryer	Master.
Thomas Ledward	Acting Surgeon.
David Nelson	Botanist.
William Peckover	Gunner.
William Cole	Boatswain.
William Purcell	Carpenter.
William Elphinston	Master's Mate.
Thomas Hayward	Midshipmen.
John Hallett	"
John Norton	Quarter Masters.
Peter Linkletter	"
Lawrence Lebogue	Sailmaker.
John Smith	Cooks.
Thomas Hall	"
George Simpson	Quarter Master's Mate.
Robert Tinkler	A boy.

Robert Lamb	Butcher.
Mr. Samuel	Clerk.

There remained on board the Bounty, as pirates,

Names.	**Stations.**
Fletcher Christian	Master's Mate.
Peter Haywood	Midshipmen.
Edward Young	"
George Stewart	"
Charles Churchill	Master at Arms.
John Mills	Gunner's Mate.
James Morrison	Boatswain's Mate.
Thomas Burkitt	Able Seaman.
Matthew Quintal	Ditto.
John Sumner	Ditto.
John Millward	Ditto.
William M'Koy	Ditto.
Henry Hillbrant	Ditto.
Michael Byrne	Ditto.
William Musprat	Ditto.
Alexander Smith	Ditto.
John Williams	Ditto.
Thomas Ellison	Ditto.

Isaac Martin	Ditto.
Richard Skinner	Ditto.
Matthew Thompson	Ditto.
William Brown	Gardiner.
Joseph Coleman	Armourer.
Charles Norman	Carpenter's Mate.
Thomas M'Intosh	Carpenter's Crew.

In all 25 hands, and the most able men of the ship's company.

1789. April.

Having little or no wind, we rowed pretty fast towards Tofoa, which bore N E about 10 leagues from us. While the ship was in sight she steered to the W N W, but I considered this only as a feint; for when we were sent away—"Huzza for Otaheite," was frequently heard among the mutineers.

Christian, the captain of the gang, is of a respectable family in the north of England. This was the third voyage he had made with me; and, as I found it necessary to keep my ship's company at three watches, I gave him an order to take charge of the third, his abilities being thoroughly equal to the task; and by this means my master and gunner were not at watch and watch.

1789. April.

Haywood is also of a respectable family in the north of England, and a young man of abilities, as well as Christian. These two were objects of my particular regard and attention, and I took great pains to instruct them, for they really promised, as professional men, to be a credit to their country.

Young was well recommended, and appeared to me an able stout seaman; therefore I was glad to take him: he, however, fell short of what his appearance promised.

Stewart was a young man of creditable parents, in the Orkneys; at which place, on the return of the Resolution from the South Seas, in 1780, we received so many civilities, that, on that account only, I should gladly have taken him with me: but, independent of this recommendation, he was a seaman, and had always borne a good character.

Notwithstanding the roughness with which I was treated, the remembrance of past kindnesses produced some signs of remorse in Christian. When they were forcing me out of the ship, I asked him, if this treatment was a proper return for the many instances he had received of my friendship? he appeared disturbed at my question, and answered, with much emotion, "That,—captain Bligh,—that is the thing;—I am in hell—I am in hell."

As soon as I had time to reflect, I felt an inward satisfaction which prevented any depression of my spirits: conscious of my integrity, and anxious solicitude for the good of the service in which I was engaged, I found my mind wonderfully supported, and I began to conceive hopes, notwithstanding so heavy a calamity, that I should one day be able to account to my King and country for the misfortune.—A few hours before, my situation had been peculiarly flattering. I had a ship in the most perfect order, and well stored with every necessary both for service and health: by early attention to those particulars I had, as much as lay in my power, provided against any accident, in case I could not get through Endeavour Straits, as well as against what might befal me in them; add to this, the plants had been successfully preserved in the most flourishing state: so that, upon the whole, the voyage was two thirds completed, and the remaining part in a very promising way; every person on board being in perfect health, to establish which was ever amongst the principal objects of my attention.

1789. April.

It will very naturally be asked, what could be the reason for such a revolt? in answer to which, I can only conjecture that the mutineers had assured themselves of a more happy life among the Otaheiteans, than they could possibly have in England; which, joined to some female connections, have most probably been the principal cause of the whole transaction.

The women at Otaheite are handsome, mild and chearful in their manners and conversation, possessed of great sensibility, and have sufficient delicacy to make them admired and beloved. The chiefs were so much attached to our people, that they rather encouraged their stay among them than otherwise, and even made them promises of large possessions. Under these, and many other attendant circumstances, equally desirable, it is now perhaps not so much to be wondered at, though scarcely possible to have been foreseen, that a set of sailors, most of them void of connections, should be led away; especially when, in addition to such powerful inducements, they imagined it in their power to fix themselves in the midst of plenty, on the finest island in the world, where they need not labour, and where the allurements of dissipation are beyond any thing that can be conceived. The utmost, however, that any commander could have supposed to have happened is, that some of the people would have been tempted to desert. But if it should be asserted, that a commander is to guard against an act of mutiny and piracy in his own ship, more than by the common rules of service, it is as much as to say that he must sleep locked up, and when awake, be girded with pistols.

1789. April.

Desertions have happened, more or less, from many of the ships that have been at the Society Islands; but it ever has been in the commanders power to make the chiefs return their people: the knowledge, therefore, that it was unsafe to desert; perhaps, first led mine to consider with what ease so small a ship might be surprized, and that so favourable an opportunity would never offer to them again.

The secrecy of this mutiny is beyond all conception. Thirteen of the party, who were with me, had always lived forward among the people; yet neither they, nor the messmates of Christian, Stewart, Haywood, and Young, had ever observed any circumstance to give them suspicion of what was going on. With such close-planned acts of villainy, and my mind free from any suspicion, it is not wonderful that I have been got the better of. Perhaps, if I had had marines, a centinel at my cabin-door might have prevented it; for I slept with the door always open, that the officer of the watch might have access to me on all occasions. The possibility of such a conspiracy was ever the farthest from my thoughts. Had their mutiny been occasioned by any grievances, either real or imaginary, I must have discovered symptoms of their discontent, which would have put me on my guard: but the case was far otherwise. Christian, in particular, I was on the most friendly terms with; that very day he was engaged to have dined with me; and the preceding night he excused himself from supping with me, on pretence of being unwell; for which I felt concerned, having no suspicions of his integrity and honour.

1789. April.

It now remained with me to consider what was best to be done. My first determination was to seek a supply of bread-fruit and water at Tofoa, and afterwards to sail for Tongataboo; and there risk a solicitation to Poulaho, the king, to equip my boat, and grant a supply of water and provisions, so as to enable us to reach the East Indies.

The quantity of provisions I found in the boat was 150 lb. of bread, 16 pieces of pork, each piece weighing 2 lb. 6 quarts of rum, 6 bottles of wine, with 28 gallons of water, and four empty barrecoes.

Wednesday 29.

Wednesday, April 29th[*]. Happily the afternoon kept calm, until about 4 o'clock, when we were so far to windward, that, with a moderate easterly breeze which sprung up, we were able to sail. It was nevertheless dark when we got to Tofoa, where I expected to land; but the shore proved to be so steep and rocky, that I was obliged to give up all thoughts of it, and keep the boat under the lee of the island with two oars; for there was no anchorage. Having fixed on this mode of proceeding for the night, I served to every person half a pint of grog, and each took to his rest as well as our unhappy situation would allow.

[*] It is to be observed, that the account of time is kept in the nautical way, each day ending at noon. Thus the beginning of the 29th of April is, according to the common way of reckoning, the afternoon of the 28th.

1789. April 29.

In the morning, at dawn of day, we set off along shore in search of landing, and about ten o'clock we discovered a stony cove at the N W part of the island, where I dropt the grapnel within 20 yards of the rocks. A great deal of surf ran on the shore; but, as I was unwilling to diminish our stock of provisions, I landed Mr. Samuel, and some others, who climbed the cliffs, and got into the country to search for supplies. The rest of us remained at the cove, not discovering any way to get into the country, but that by which Mr. Samuel had proceeded. It was great consolation to me to find, that the spirits of my people did not sink, notwithstanding our miserable and almost hopeless situation. Towards noon Mr. Samuel returned, with a few quarts of water, which he had found in holes; but he had met with no spring or any prospect of a sufficient supply in that particular, and had only seen signs of inhabitants. As it was impossible to know how much we might be in want, I only issued a morsel of bread, and a glass of wine, to each person for dinner.

I observed the latitude of this cove to be 19° 41´ S.

This is the N W part of Tofoa, the north-westernmost of the Friendly Islands.

Thursday 30.

Thursday, April 30th. Fair weather, but the wind blew so violently from the E S E that I could not venture to sea. Our detention therefore made it absolutely necessary to see what we could do more for our support; for I determined, if possible, to keep my first stock entire: I therefore weighed, and rowed along shore, to see if any thing could be got; and at last discovered some cocoa-nut trees, but they were on the top of high precipices, and the surf made it dangerous landing; both one and the other we, however, got the better of. Some, with much difficulty, climbed the cliffs, and got about 20 cocoa-nuts, and others slung them to ropes, by which we hauled them through the surf into the boat. This was all that could be done here; and, as I found no place so eligible as the one we had left to spend the night at, I returned to the cove, and, having served a cocoa-nut to each person, we went to rest again in the boat.

1789. April 30.

At dawn of day I attempted to get to sea; but the wind and weather proved so bad, that I was glad to return to my former station; where, after issuing a morsel of bread and a spoonful of rum to each person, we landed, and I went off with Mr. Nelson, Mr. Samuel, and some others, into the country, having hauled ourselves up the precipice by long vines, which were fixed there by the natives for that purpose; this being the only way into the country.

We found a few deserted huts, and a small plantain walk, but little taken care of; from which we could only collect three small bunches of plantains. After passing this place, we came to a deep gully that led towards a mountain, near a volcano; and, as I conceived that in the rainy season very great torrents of water must pass through it, we hoped to find sufficient for our use remaining in some holes of the rocks; but, after all our search, the whole that we found was only nine gallons, in the course of the day. We advanced within two miles of the foot of the highest mountain in the island, on which is the volcano that is almost constantly burning. The country near it is all covered with lava, and has a most dreary appearance. As we had not been fortunate in our discoveries, and saw but little to alleviate our distresses, we filled our cocoa-nut shells with the water we found, and returned exceedingly fatigued and faint. When I came to the precipice whence we were to descend into the cove, I was seized with such a dizziness in my head, that I thought it scarce possible to effect it: however, by the assistance of Mr. Nelson, and others, they at last got me down, in a weak condition. Every person being returned by noon, I gave about an ounce of pork and two plantains to each, with half a glass of wine. I again observed the latitude of this place 19° 41´ south. The people who remained by the boat I had directed to look for fish, or what they could pick up about the rocks; but nothing eatable could be found: so that, upon the whole, we considered ourselves on as miserable a spot of land as could well be imagined.

I could not say positively, from the former knowledge I had of this island, whether it was inhabited or not; but I knew it was considered inferior to the other islands, and I was not certain but that the Indians only resorted to it at particular times. I was very anxious to ascertain this point; for, in case there had only been a few people here, and those could have furnished us with but very moderate supplies, the remaining in this spot to have made preparations for our voyage, would have been preferable to the risk of going amongst multitudes, where perhaps we might lose every thing. A party, therefore, sufficiently strong, I determined should go another route, as soon as the sun became lower; and they cheerfully undertook it.

May. Friday 1.

Friday, May the 1st: stormy weather, wind E S E and S E. About two o'clock in the afternoon the party set out; but, after suffering much fatigue, they returned in the evening, without any kind of success.

At the head of the cove, about 150 yards from the water-side, was a cave; across the stony beach was about 100 yards, and the only way from the country into the cove was that which I have already described. The situation secured us from the danger of being surprised, and I determined to remain on shore for the night, with a part of my people, that the others might have more room to rest in the boat, with the master; whom I directed to lie at a grapnel, and be watchful, in case we should be attacked. I ordered one plantain for each person to be boiled; and, having supped on this scanty allowance, with a quarter of a pint of grog, and fixed the watches for the night, those whose turn it was, laid down to sleep in the cave; before which we kept up a good fire, yet notwithstanding we were much troubled with flies and musquitoes.

MAY

1789. May 1.

At dawn of day the party set out again in a different route, to see what they could find; in the course of which they suffered greatly for want of water: they, however, met with two men, a woman, and a child; the men came with them to the cove, and brought two cocoa-nut shells of water. I immediately made friends with these people, and sent them away for bread-fruit, plantains, and water. Soon after other natives came to us; and by noon I had 30 of them about me, trading with the articles we were in want of: but I could only afford one ounce of pork, and a quarter of a bread-fruit, to each man for dinner, with half a pint of water; for I was fixed in not using any of the bread or water in the boat.

No particular chief was yet among the natives: they were, notwithstanding, tractable, and behaved honestly, giving the provisions they brought for a few buttons and beads. The party who had been out, informed me of having discovered several neat plantations; so that it became no longer a doubt of there being settled inhabitants on the island; and for that reason I determined to get what I could, and sail the first moment the wind and weather would allow me to put to sea.

1789. May 1.

Saturday 2.

Saturday, May the 2d: stormy weather, wind E S E. It had hitherto been a weighty consideration with me, how I was to account to the natives for the loss of my ship: I knew they had too much sense to be amused with a story that the ship was to join me, when she was not in sight from the hills. I was at first doubtful whether I should tell the real fact, or say that the ship had overset and sunk, and that only we were saved: the latter appeared to me to be the most proper and advantageous to us, and I accordingly instructed my people, that we might all agree in one story. As I expected, enquiries were made after the ship, and they seemed readily satisfied with our account; but there did not appear the least symptom of joy or sorrow in their faces, although I fancied I discovered some marks of surprise. Some of the natives were coming and going the whole afternoon, and we got enough of bread-fruit, plantains, and cocoa-nuts for another day; but water they only brought us about five pints. A canoe also came in with four men, and brought a few cocoa-nuts and bread-fruit, which I bought as I had done the rest. Nails were much enquired after, but I would not suffer one to be shewn, as I wanted them for the use of the boat.

Towards evening I had the satisfaction to find our stock of provisions somewhat increased: but the natives did not appear to have much to spare. What they brought was in such small quantities, that I had no reason to hope we should be able to procure from them sufficient to stock us for our voyage. At sun-set all the natives left us in quiet possession of the cove. I thought this a good sign, and made no doubt that they would come again the next day with a larger proportion of food and water, with which I hoped to sail without farther delay: for if, in attempting to get to Tongataboo, we should be blown away from the islands altogether, there would be a larger quantity of provisions to support us against such a misfortune.

1789. May 2.

At night I served a quarter of a bread-fruit and a cocoa-nut to each person for supper; and, a good fire being made, all but the watch went to sleep.

At day-break I was happy to find every one's spirits a little revived, and that they no longer regarded me with those anxious looks, which had constantly been directed towards me since we lost sight of the ship: every countenance appeared to have a degree of cheerfulness, and they all seemed determined to do their best.

As I doubted of water being brought by the natives, I sent a party among the gullies in the mountains, with empty shells, to see what they could get. In their absence the natives came about us, as I expected, but more numerous; also two canoes came in from round the north side of the island. In one of them was an elderly chief, called Maccaackavow. Soon after some of our foraging party returned, and with them came a good-looking chief, called Eegijeefow, or perhaps more properly Eefow, Egij or Eghee, signifying a chief. To both these men I made a present of an old shirt and a knife, and I soon found they either had seen me, or had heard of my being at Annamooka. They knew I had been with captain Cook, who they enquired after, and also captain Clerk. They were very inquisitive to know in what manner I had lost my ship. During this conversation a young man appeared, whom I remembered to have seen at Annamooka, called Nageete: he expressed much pleasure at seeing me. I now enquired after Poulaho and Feenow, who, they said, were at Tongataboo; and Eefow agreed to accompany me thither, if I would wait till the weather moderated. The readiness and affability of this man gave me much satisfaction.

1789. May 2.

This, however, was but of short duration, for the natives began to increase in number, and I observed some symptoms of a design against us; soon after they attempted to haul the boat on shore, when I threatened Eefow with a cutlass, to induce him to make them desist; which they did, and every thing became quiet again. My people, who had been in the mountains, now returned with about three gallons of water. I kept buying up the little bread-fruit that was brought to us, and likewise some spears to arm my men with, having only four cutlasses, two of which were in the boat. As we had no means of improving our situation, I told our people I would wait until sun-set, by which time, perhaps, something might happen in our favour: that if we attempted to go at present, we must fight our way through, which we could do more advantageously at night; and that in the mean time we would endeavour to get off to the boat what we had bought. The beach was now lined with the natives, and we heard nothing but the knocking of stones together, which they had in each hand. I knew very well this was the sign of an attack. It being now noon, I served a cocoa-nut and a bread-fruit to each person for dinner, and gave some to the chiefs, with whom I continued to appear intimate and friendly. They frequently importuned me to sit down, but I as constantly refused; for it occurred both to Mr. Nelson and myself, that they intended to seize hold of me, if I gave them such an opportunity. Keeping, therefore, constantly on our guard, we were suffered to eat our uncomfortable meal in some quietness.

1789. May 2.

Sunday 3.

Sunday, 3d May, fresh gales at S E and E S E, varying to the N E in the latter part, with a storm of wind.

After dinner we began by little and little to get our things into the boat, which was a troublesome business, on account of the surf. I carefully watched the motions of the natives, who still increased in number, and found that, instead of their intention being to leave us, fires were made, and places fixed on for their stay during the night. Consultations were also held among them, and every thing assured me we should be attacked. I sent orders to the master, that when he saw us coming down, he should keep the boat close to the shore, that we might the more readily embark.

I had my journal on shore with me, writing the occurrences in the cave, and in sending it down to the boat it was nearly snatched away, but for the timely assistance of the gunner.

The sun was near setting when I gave the word, on which every person, who was on shore with me, boldly took up his proportion of things, and carried them to the boat. The chiefs asked me if I would not stay with them all night, I said, "No, I never sleep out of my boat; but in the morning we will again trade with you, and I shall remain until the weather is moderate, that we may go, as we have agreed, to see Poulaho, at Tongataboo." Maccaackavow then got up, and said, "You will not sleep on shore? then Mattie," (which directly signifies we will kill you) and he left me. The onset was now preparing; every one, as I have described before, kept knocking stones together, and Eefow quitted me. We had now all but two or three things in the boat, when I took Nageete by the hand, and we walked down the beach, every one in a silent kind of horror.

1789. May 3.

When I came to the boat, and was seeing the people embark, Nageete wanted me to stay to speak to Eefow; but I found he was encouraging them to the attack, and I determined, had it then begun, to have killed him for his treacherous behaviour. I ordered the carpenter not to quit me until the other people were in the boat. Nageete, finding I would not stay, loosed himself from my hold and went off, and we all got into the boat except one man, who, while I was getting on board, quitted it, and ran up the beach to cast the stern fast off, notwithstanding the master and others called to him to return, while they were hauling me out of the water.

I was no sooner in the boat than the attack began by about 200 men; the unfortunate poor man who had run up the beach was knocked down, and the stones flew like a shower of shot. Many Indians got hold of the stern rope, and were near hauling us on shore, and would certainly have done it if I had not had a knife in my pocket, with which I cut the rope. We then hauled off to the grapnel, every one being more or less hurt. At this time I saw five of the natives about the poor man they had killed, and two of them were beating him about the head with stones in their hands.

1789. May 3

We had no time to reflect, before, to my surprise, they filled their canoes with stones, and twelve men came off after us to renew the attack, which they did so effectually as nearly to disable all of us. Our grapnel was foul, but Providence here assisted us; the fluke broke, and we got to our oars, and pulled to sea. They, however, could paddle round us, so that we were obliged to sustain the attack without being able to return it, except with such stones as lodged in the boat, and in this I found we were very inferior to them. We could not close, because our boat was lumbered and heavy, and that they knew very well: I therefore adopted the expedient of throwing overboard some cloaths, which they lost time in picking up; and, as it was now almost dark, they gave over the attack, and returned towards the shore, leaving us to reflect on our unhappy situation.

The poor man I lost was John Norton: this was his second voyage with me as a quarter-master, and his worthy character made me lament his loss very much. He has left an aged parent, I am told, whom he supported.

1789. May 3.

I once before sustained an attack of a similar nature, with a smaller number of Europeans, against a multitude of Indians; it was after the death of captain Cook, on the Morai at Owhyhee, where I was left by lieutenant King: yet, notwithstanding, I did not conceive that the power of a man's arm could throw stones, from two to eight pounds weight, with such force and exactness as these people did. Here unhappily I was without arms, and the Indians knew it; but it was a fortunate circumstance that they did not begin to attack us in the cave: in that case our destruction must have been inevitable, and we should have had nothing left for it but to die as bravely as we could, fighting close together; in which I found every one cheerfully disposed to join me. This appearance of resolution deterred them, supposing they could effect their purpose without risk after we were in the boat.

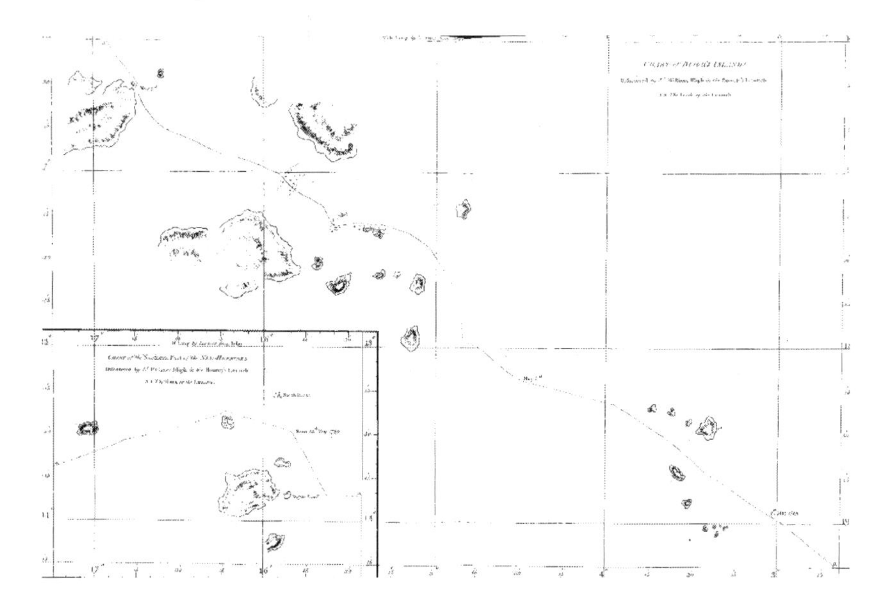

Chart of Bligh's Islands

Taking this as a sample of the dispositions of the Indians, there was little reason to expect much benefit if I persevered in my intention of visiting Poulaho; for I considered their good behaviour hitherto to proceed from a dread of our fire-arms, which, now knowing us destitute of, would cease; and, even supposing our lives not in danger, the boat and every thing we had would most probably be taken from us, and thereby all hopes precluded of ever being able to return to our native country.

We were now sailing along the west side of the island Tofoa, and my mind was employed in considering what was best to be done, when I was solicited by all hands to take them towards home: and, when I told them no hopes of relief for us remained, but what I might find at New Holland, until I came to Timor, a distance of full 1200 leagues, where was a Dutch settlement, but in what part of the island I knew not, they all agreed to live on one ounce of bread, and a quarter of a pint of water, per day. Therefore, after examining our stock of provisions, and recommending this as a sacred promise for ever to their memory, we bore away across a sea, where the navigation is but little known, in a small boat, twenty-three feet long from stern to stern, deep laden with eighteen men; without a chart, and nothing but my own recollection and general knowledge of the situation of places, assisted by a book of latitudes and longitudes, to guide us. I was happy, however, to see every one better satisfied with our situation in this particular than myself.

1789. May 3.

Our stock of provisions consisted of about one hundred and fifty pounds of bread, twenty-eight gallons of water, twenty pounds of pork, three bottles of wine, and five quarts of rum. The difference between this and the quantity we had on leaving the ship, was principally owing to loss in the bustle and confusion of the attack. A few cocoa-nuts were in the boat, and some bread-fruit, but the latter was trampled to pieces.

It was about eight o'clock at night when I bore away under a reefed lug fore-sail: and, having divided the people into watches, and got the boat in a little order, we returned God thanks for our miraculous preservation, and, fully confident of his gracious support, I found my mind more at ease than for some time past.

At day-break the gale increased; the sun rose very fiery and red, a sure indication of a severe gale of wind. At eight it blew a violent storm, and the sea ran very high, so that between the seas the sail was becalmed, and when on the top of the sea it was too much to have set: but I was obliged to carry to it, for we were now in very imminent danger and distress, the sea curling over the stern of the boat, which obliged us to bale with all our might. A situation more distressing has, perhaps, seldom been experienced.

Our bread was in bags, and in danger of being spoiled by the wet: to be starved to death was inevitable, if this could not be prevented: I therefore began to examine what cloaths there were in the boat, and what other things could be spared; and, having determined that only two suits should be kept for each person, the rest was thrown overboard, with some rope and spare sails, which lightened the boat considerably, and we had more room to bale the water out. Fortunately the carpenter had a good chest in the boat, into which I put the bread the first favourable moment. His tool chest also was cleared, and the tools stowed in the bottom of the boat, so that this became a second convenience.

1789. May 3.

I now served a tea-spoonful of rum to each person, (for we were very wet and cold) with a quarter of a bread-fruit, which was scarce eatable, for dinner; but our engagement was now strictly to be carried into execution, and I was fully determined to make what provisions I had last eight weeks, let the daily proportion be ever so small.

At noon I considered my course and distance from Tofoa to be W N W 3/4 W. 86 miles, my latitude 19° 27´ S. I directed my course to the W N W, that I might get a sight of the islands called Feejee, if they laid in the direction the natives had pointed out to me.

Monday 4.

Monday, 4th May. This day the weather was very severe, it blew a storm from N E to E S E. The sea ran higher than yesterday, and the fatigue of baling, to keep the boat from filling, was exceedingly great. We could do nothing more than keep before the sea; in the course of which the boat performed so wonderfully well, that I no longer dreaded any danger in that respect. But among the hardships we were to undergo, that of being constantly wet was not the least: the nights were very cold, and at day-light our limbs were so benumbed, that we could scarce find the use of them. At this time I served a tea-spoonful of rum to each person, which we all found great benefit from.

1789. May 4.

As I have mentioned before, I determined to keep to the W N W, until I got more to the northward, for I not only expected to have better weather, but to see the Feejee Islands, as I have often understood, from the natives of Annamooka, that they lie in that direction; Captain Cook likewise considers them to be N W by W from Tongataboo. Just before noon we discovered a small flat island of a moderate height, bearing W S W, 4 or 5 leagues. I observed in latitude 18° 58′ S; our longitude, by account, 3° 4′ W from the island Tofoa, having made a N 72° W course, distance 95 miles, since yesterday noon. I divided five small cocoa-nuts for our dinner, and every one was satisfied.

Tuesday 5.

Tuesday, 5th May. Towards the evening the gale considerably abated. Wind S E.

A little after noon, other islands appeared, and at a quarter past three o'clock we could count eight, bearing from S round by the west to N W by N; those to the south, which were the nearest, being four leagues distant from us.

I kept my course to the N W by W, between the islands, and at six o'clock discovered three other small islands to the N W, the westernmost of them bore N W 1/2 W 7 leagues. I steered to the southward of these islands, a W N W course for the night, under a reefed sail.

Served a few broken pieces of bread-fruit for supper, and performed prayers.

The night turned out fair, and, having had tolerable rest, every one seemed considerably better in the morning, and contentedly breakfasted on a few pieces of yams that were found in the boat. After breakfast we prepared a chest for our bread, and it got secured: but unfortunately a great deal was damaged and rotten; this nevertheless we were glad to keep for use.

I had hitherto been scarcely able to keep any account of our run; but we now equipped ourselves a little better, by getting a log-line marked, and, having practised at counting seconds; several could do it with some degree of exactness.

1789. May 5.

The islands I have passed lie between the latitude of 19° 5´ S and 18° 19´ S, and, according to my reckoning, from 3° 17´ to 3° 46´ W longitude from the island Tofoa: the largest may be about six leagues in circuit; but it is impossible for me to be very exact. To show where they are to be found again is the most my situation enabled me to do. The sketch I have made, will give a comparative view of their extent. I believe all the larger islands are inhabited, as they appeared very fertile.

At noon I observed, in latitude 18° 10´ S, and considered my course and distance from yesterday noon, N W by W 1/2 W, 94 miles; longitude, by account, from Tofoa 4° 29´ W.

For dinner, I served some of the damaged bread, and a quarter of a pint of water.

Wednesday 6.

Wednesday, 6th May. Fresh breezes E N E, and fair weather, but very hazy.

About six o'clock this afternoon I discovered two islands, one bearing W by S 6 leagues, and the other N W by N 8 leagues; I kept to windward of the northernmost, and passing it by 10 o'clock, I resumed my course to the N W and W N W. At day-light in the morning I discovered a number of other islands from S S E to the W, and round to N E by E; between those in the N W I determined to pass. At noon a small sandy island or key, 2 miles distant from me, bore from E to S 3/4 W. I had passed ten islands, the largest of which may be 6 or 8 leagues in circuit. Much larger lands appeared in the S W and N by W, between which I directed my course. Latitude observed 17° 17′ S; course since yesterday noon N 50° W; distance 84 miles; longitude made, by account, 5° 37′ W.

1789. May 6.

Our supper, breakfast, and dinner, consisted of a quarter of a pint of cocoa-nut milk, and the meat, which did not exceed two ounces to each person: it was received very contentedly, but we suffered great drought. I dared not to land, as we had no arms, and were less capable to defend ourselves than we were at Tofoa.

To keep an account of the boat's run was rendered difficult, from being constantly wet with the sea breaking over us; but, as we advanced towards the land, the sea became smoother, and I was enabled to form a sketch of the islands, which will serve to give a general knowledge of their extent. Those I have been near are fruitful and hilly, some very mountainous, and all of a good height.

To our great joy we hooked a fish, but we were miserably disappointed by its being lost in getting into the boat.

Thursday 7.

Thursday, 7th May. Variable weather and cloudy, wind north-easterly, and calms. I continued my course to the N W, between the islands, which, by the evening, appeared of considerable extent, woody and mountainous. At sun-set the southernmost bore from S to S W by W, and the northernmost from N by W 1/2 W to N E 1/2 E. At six o'clock I was nearly mid-way between them, and about 6 leagues distant from each shore, when I fell in with a coral bank, where I had only four feet water, without the least break on it, or ruffle of the sea to give us warning. I could only see that it extended about a mile on each side of us; but, as it is probable that it extends much farther, I have laid it down so in my sketch.

I now directed my course W by N for the night, and served to each person an ounce of the damaged bread, and a quarter of a pint of water, for supper.

1789. May 7.

It may readily be supposed, that our lodgings were very miserable and confined, and I had only in my power to remedy the latter defect by putting ourselves at watch and watch; so that one half always sat up while the other lay down on the boat's bottom, or upon a chest, with nothing to cover us but the heavens. Our limbs were dreadfully cramped, for we could not stretch them out, and the nights were so cold, and we so constantly wet, that after a few hours sleep we could scarce move.

At dawn of day we again discovered land from W S W to W N W, and another island N N W, the latter a high round lump of but little extent; and I could see the southern land that I had passed in the night. Being very wet and cold, I served a spoonful of rum and a morsel of bread for breakfast.

1789. May 7.

As I advanced towards the land in the west, it appeared in a variety of forms; some extraordinary high rocks, and the country agreeably interspersed with high and low land, covered in some places with wood. Off the N E part lay two small rocky islands, between which and the island to the N E, 4 leagues apart, I directed my course; but a lee current very unexpectedly set us very near to the shore, and I could only get clear of it by rowing, passing close to the reef that surrounded the rocky isles. We now observed two large sailing canoes coming swiftly after us along shore, and, being apprehensive of their intentions, we rowed with some anxiety, being sensible of our weak and defenceless state. It was now noon, calm and cloudy weather, my latitude is therefore doubtful to 3 or 4 miles; my course since yesterday noon N 56 W, distance 79 miles; latitude by account, 16° 29´ S, and longitude by account, from Tofoa, 6° 46´ W. Being constantly wet, it was with the utmost difficulty I could open a book to write, and I am sensible that what I have done can only serve to point out where these lands are to be found again, and give an idea of their extent.

Friday 8.

Friday, 8th May. All the afternoon the weather was very rainy, attended with thunder and lightning. Wind N N E.

Only one of the canoes gained upon us, and by three o'clock in the afternoon was not more than two miles off, when she gave over chase.

If I may judge from the sail of the vessels, they are the same as at the Friendly Islands, and the nearness of their situation leaves little room to doubt of their being the same kind of people. Whether these canoes had any hostile intention against us is a matter of doubt; perhaps we might have benefited by an intercourse with them, but in our defenceless situation it would have been risking too much to make the experiment.

I imagine these to be the islands called Feejee, as their extent, direction, and distance from the Friendly Islands, answers to the description given of them by those Islanders. Heavy rain came on at four o'clock, when every person did their utmost to catch some water, and we increased our stock to 34 gallons, besides quenching our thirst for the first time since we had been at sea; but an attendant consequence made us pass the night very miserably, for, being extremely wet, and no dry things to shift or cover us, we experienced cold and shiverings scarce to be conceived. Most fortunately for us, the forenoon turned out fair, and we stripped and dried our cloaths. The allowance I issued to-day, was an ounce and a half of pork, a tea-spoonful of rum, half a pint of cocoa-nut milk, and an ounce of bread. The rum, though so small in quantity, was of the greatest service. A fishing-line was generally towing, and we saw great numbers of fish, but could never catch one.

At noon, I observed, in latitude 16° 4´ S, and found I had made a course, from yesterday noon, N 62° W, distance 62 miles; longitude, by account, from Tofoa, 7° 42´ W.

1789. May 8.

The land I passed yesterday, and the day before, is a group of islands, 14 or 16 in number, lying between the latitude of 16° 26′ S and 17° 57′ S, and in longitude, by my account, 4° 47′ to 7° 17′ W from Tofoa; three of these islands are very large, having from 30 to 40 leagues of sea-coast.

Saturday 9.

Saturday, 9th May. Fine weather, and light winds from the N E to E by S.

This afternoon we cleaned out the boat, and it employed us till sun-set to get every thing dry and in order. Hitherto I had issued the allowance by guess, but I now got a pair of scales, made with two cocoa-nut shells; and, having accidentally some pistol-balls in the boat, 25[*] of which weighed one pound, or 16 ounces, I adopted one, as the proportion of weight that each person should receive of bread at the times I served it. I also amused all hands, with describing the situation of New Guinea and New Holland, and gave them every information in my power, that in case any accident happened to me, those who survived might have some idea of what they were about, and be able to find their way to Timor, which at present they knew nothing of, more than the name, and some not that.

[*] It weighed 272 grains.

At night I served a quarter of a pint of water, and half an ounce of bread, for supper. In the morning, a quarter of a pint of cocoa-nut milk, and some of the decayed bread, for breakfast; and for dinner, I divided the meat of four cocoa-nuts, with the remainder of the rotten bread, which was only eatable by such distressed people.

At noon, I observed the latitude to be 15° 47′ S; course since yesterday N 75° W; distant 64 miles; longitude made, by account, 8° 45′ W.

1789 May 10.

Sunday 10.

Sunday, May the 10th. The first part of this day fine weather; but after sun-set it became squally, with hard rain, thunder, and lightning, and a fresh gale; wind E by S, S E, and S S E.

In the afternoon I got fitted a pair of shrouds for each mast and contrived a canvass weather cloth round the boat, and raised the quarters about nine inches, by nailing on the seats of the stern sheets, which proved of great benefit to us.

About nine o'clock in the evening, the clouds began to gather, and we had a prodigious fall of rain, with severe thunder and lightning. By midnight we had caught about twenty gallons of water. Being miserably wet and cold, I served to each person a tea-spoonful of rum, to enable them to bear with their distressed situation. The weather continued extremely bad, and the wind increased; we spent a very miserable night, without sleep, but such as could be got in the midst of rain. The day brought us no relief but its light. The sea was constantly breaking over us, which kept two persons baling; and we had no choice how to steer, for we were obliged to keep before the waves to avoid filling the boat.

The allowance which I now regularly served to each person was one 25th of a pound of bread, and a quarter of a pint of water, at sun-set, eight in the morning, and at noon. To-day I gave about half an ounce of pork for dinner, which, though any moderate person would have considered but a mouthful, was divided into three or four.

The rain abated towards noon, and I observed the latitude to be 15° 17´ S; course N 67° W; distance 78 miles; longitude made 10° W.

Monday 11.

Monday, May the 11th. Strong gales from S S E to S E, and very squally weather, with a high breaking sea, so that we were miserably wet, and suffered great cold in the night. In the morning at day-break I served to every person a tea-spoonful of rum, our limbs being so cramped that we could scarce feel the use of them. Our situation was now extremely dangerous, the sea frequently running over our stern, which kept us baling with all our strength.

1789. May 11.

At noon the sun appeared, which gave us as much pleasure as in a winter's day in England. I issued the 25th of a pound of bread, and a quarter of a pint of water, as yesterday. Latitude observed 14° 50′ S; course N 71° W; distance 102 miles; and longitude, by account, 11° 39′ W. from Tofoa.

Tuesday 12.

Tuesday, May the 12th. Strong gales at S E, with much rain and dark dismal weather, moderating towards noon and wind varying to the N E.

Having again experienced a dreadful night, the day showed to me a poor miserable set of beings full of wants, without any thing to relieve them. Some complained of a great pain in their bowels, and all of having but very little use of their limbs. What sleep we got was scarce refreshing, we being covered with sea and rain. Two persons were obliged to be always baling the water out of the boat. I served a spoonful of rum at day-dawn, and the usual allowance of bread and water, for supper, breakfast, and dinner.

At noon it was almost calm, no sun to be seen, and some of us shivering with cold. Course since yesterday W by N; distance 89 miles; latitude, by account, 14° 33′ S; longitude made 13° 9′ W. The direction of my course is to pass to the northward of the New Hebrides.

Wednesday 13.

Wednesday, May the 13th. Very squally weather, wind southerly. As I saw no prospect of getting our cloaths dried, I recommended it to every one to strip, and wring them through the salt water, by which means they received a warmth, that, while wet with rain, they could not have, and we were less liable to suffer from colds or rheumatic complaints.

1789. May 13.

In the afternoon we saw a kind of fruit on the water, which Mr. Nelson knew to be the Barringtonia of Forster, and, as I saw the same again in the morning, and some men of war birds, I was led to believe we were not far from land.

We continued constantly shipping seas, and baling, and were very wet and cold in the night; but I could not afford the allowance of rum at day-break. The twenty-fifth of a pound of bread, and water I served as usual. At noon I had a sight of the sun, latitude 14° 17´ S; course W by N 79 miles; longitude made 14° 28´ W.

Thursday 14.

Thursday, May the 14th. Fresh breezes and cloudy weather, wind southerly. Constantly shipping water, and very wet, suffering much cold and shiverings in the night. Served the usual allowance of bread and water, three times a day.

At six in the morning, we saw land, from S W by S eight leagues, to N W by W 3/4 W six leagues, which I soon after found to be four islands, all of them high and remarkable. At noon discovered a rocky island N W by N four leagues, and another island W eight leagues, so that the whole were six in number; the four I had first seen bearing from S 1/2 E to S W by S; our distance three leagues from the nearest island. My latitude observed was 13° 29´ S, and longitude, by account, from Tofoa, 15° 49´ W; course since yesterday noon N 63° W; distance 89 miles.

Friday 15.

Friday, May the 15th. Fresh gales at S E, and gloomy weather with rain, and a very high sea; two people constantly employed baling.

1789. May 15.

At four in the afternoon I passed the westernmost island. At one in the morning I discovered another, bearing W N W, five leagues distance, and at eight o'clock I saw it for the last time, bearing N E seven leagues. A number of gannets, boobies, and men of war birds were seen.

These islands lie between the latitude of 13° 16´ S and 14° 10´ S: their longitude, according to my reckoning, 15° 51´ to 17° 6´ W from the island Tofoa[*]. The largest island may be twenty leagues in circuit, the others five or six. The easternmost is the smallest island, and most remarkable, having a high sugar-loaf hill.

[*] By making a proportional allowance for the error afterwards found in the dead reckoning, I estimate the longitude of these islands to be from 167° 17´ E to 168° 34´ E from Greenwich.

The sight of these islands served but to increase the misery of our situation. We were very little better than starving, with plenty in view; yet to attempt procuring any relief was attended with so much danger, that prolonging of life, even in the midst of misery, was thought preferable, while there remained hopes of being able to surmount our hardships. For my own part, I consider the general run of cloudy and wet weather to be a blessing of Providence. Hot weather would have caused us to have died with thirst; and perhaps being so constantly covered with rain or sea protected us from that dreadful calamity.

As I had nothing to assist my memory, I could not determine whether these islands were a part of the New Hebrides or not: I believed them perfectly a new discovery, which I have since found to be the case; but, though they were not seen either by Monsieur Bougainville or Captain Cook, they are so nearly in the neighbourhood of the New Hebrides, that they must be considered as part of the same group. They are fertile, and inhabited, as I saw smoke in several places.

1789. May 16.

Saturday 16.

Saturday, May the 16th. Fresh gales from the S E, and rainy weather. The night was very dark, not a star to be seen to steer by, and the sea breaking constantly over us. I found it necessary to act as much as possible against the southerly winds, to prevent being driven too near

New Guinea; for in general we were forced to keep so much before the sea, that if we had not, at intervals of moderate weather, steered a more southerly course, we should inevitably, from a continuance of the gales, have been thrown in sight of that coast: in which case there would most probably have been an end to our voyage.

In addition to our miserable allowance of one 25th of a pound of bread, and a quarter of a pint of water, I issued for dinner about an ounce of salt pork to each person. I was often solicited for this pork, but I considered it better to give it in small quantities than to use all at once or twice, which would have been done if I had allowed it.

At noon I observed, in 13° 33′ S; longitude made from Tofoa, 19° 27′ W; course N 82° W; distance 101 miles. The sun gave us hopes of drying our wet cloaths.

Sunday 17.

Sunday, May the 17th. The sunshine was but of short duration. We had strong breezes at S E by S, and dark gloomy weather, with storms of thunder, lightning, and rain. The night was truly horrible, and not a star to be seen; so that our steerage was uncertain. At dawn of day I found every person complaining, and some of them soliciting extra allowance; but I positively refused it. Our situation was extremely miserable; always wet, and suffering extreme cold in the night, without the least shelter from the weather. Being constantly obliged to bale, to keep the boat from filling, was, perhaps, not to be reckoned an evil, as it gave us exercise.

1789. May 17.

The little rum I had was of great service to us; when our nights were particularly distressing, I generally served a tea-spoonful or two to each person: and it was always joyful tidings when they heard of my intentions.

At noon a water-spout was very near on board of us. I issued an ounce of pork, in addition to the allowance of bread and water; but before we began to eat, every person stript and wrung their cloaths through the sea-water, which we found warm and refreshing. Course since yesterday noon W S W; distance 100 miles; latitude, by account, 14° 11´ S, and longitude made 21° 3´ W.

Monday 18.

Monday, May the 18th. Fresh gales with rain, and a dark dismal night, wind S E; the sea constantly breaking over us, and nothing but the wind and sea to direct our steerage. I now fully determined to make New Holland, to the southward of Endeavour straits, sensible that it was necessary to preserve such a situation as would make a southerly wind a fair one; that I might range the reefs until an opening should be found into smooth water, and we the sooner be able to pick up some refreshments.

In the morning the rain abated, when we stripped, and wrung our cloaths through the sea-water, as usual, which refreshed us wonderfully. Every person complained of violent pain in their bones: I was only surprised that no one was yet laid up. Served one 25th of a pound of bread, and a quarter of a pint of water, at supper, breakfast, and dinner, as customary.

At noon I deduced my situation, by account, for we had no glimpse of the sun, to be in latitude 14° 52´ S; course since yesterday noon W S W 106 miles; longitude made from Tofoa 22° 45´ W. Saw many boobies and noddies, a sign of being in the neighbourhood of land.

1789. May 19. Tuesday 19.

Tuesday, May the 19th. Fresh gales at E N E, with heavy rain, and dark gloomy weather, and no sight of the sun. We past this day miserably wet and cold, covered with rain and sea, from which we had no relief, but at intervals by pulling off our cloaths and wringing them through the sea water. In the night we had very severe lightning, but otherwise it was so dark that we could not see each other. The morning produced many complaints on the severity of the weather, and I would gladly have issued my allowance of rum, if it had not appeared to me that we were to suffer much more, and that it was necessary to preserve the little I had, to give relief at a time we might be less able to bear such hardships; but, to make up for it, I served out about half an ounce of pork to each person, with the common allowance of bread and water, for dinner. All night and day we were obliged to bale without intermission.

At noon it was very bad weather and constant rain; latitude, by account, 14° 37´ S; course since yesterday N 81° W; distance 100 miles; longitude made 24° 30´ W.

Wednesday 20.

Wednesday, May the 20th. Fresh breezes E N E with constant rain; at times a deluge. Always baling.

1789. May 20.

At dawn of day, some of my people seemed half dead: our appearances were horrible; and I could look no way, but I caught the eye of some one in distress. Extreme hunger was now too evident, but no one suffered from thirst, nor had we much inclination to drink, that desire, perhaps, being satisfied through the skin. The little sleep we got was in the midst of water, and we constantly awoke with severe cramps and pains in our bones. This morning I served about two tea-spoonfuls of rum to each person, and the allowance of bread and water, as usual. At noon the sun broke out, and revived every one. I found we were in latitude 14° 49´ S; longitude made 25° 46´ W; course S 88° W; distance 75 miles.

Thursday 21.

Thursday, May the 21st. Fresh gales, and heavy showers of rain. Wind E N E.

Our distresses were now very great, and we were so covered with rain and salt water, that we could scarcely see. Sleep, though we longed for it, afforded no comfort: for my own part, I almost lived without it: we suffered extreme cold, and every one dreaded the approach of night. About two o'clock in the morning we were overwhelmed with a deluge of rain. It fell so heavy that we were afraid it would fill the boat, and were obliged to bale with all our might. At dawn of day, I served a large allowance of rum. Towards noon the rain abated and the sun shone, but we were miserably cold and wet, the sea breaking so constantly over us, that, notwithstanding the heavy rain, we had not been able to add to our stock of fresh water. The usual allowance of one 25th of a pound of bread and water was served at evening, morning, and noon. Latitude, by observation, 14° 29´ S, and longitude made, by account, from Tofoa, 27° 25´ W; course, since yesterday noon, N 78° W, 99 miles. I now considered myself on a meridian with the east part of New Guinea, and about 65 leagues distant from the coast of New Holland.

Friday 22.

Friday, May the 22nd. Strong gales from E S E to S S E, a high sea, and dark dismal night.

Our situation this day was extremely calamitous. We were obliged to take the course of the sea, running right before it, and watching with the utmost care, as the least error in the helm would in a moment have been our destruction. The sea was continually breaking all over us; but, as we suffered not such cold as when wet with the rain, I only served the common allowance of bread and water.

1789. May 22.

At noon it blew very hard, and the foam of the sea kept running over our stern and quarters; I however got propped up, and made an observation of the latitude, in 14° 17´ S; course N 85° W; distance 130 miles; longitude made 29° 38´ west.

Saturday 23.

Saturday, May the 23d. Strong gales with very hard squalls, and rain; wind S E, and S S E.

The misery we suffered this day exceeded the preceding. The night was dreadful. The sea flew over us with great force, and kept us baling with horror and anxiety. At dawn of day I found every one in a most distressed condition, and I now began to fear that another such a night would put an end to the lives of several who seemed no longer able to support such sufferings. Every one complained of severe pains in their bones; but these were alleviated, in some degree, by an allowance of two tea-spoonfuls of rum; after drinking which, having wrung our cloaths, and taken our breakfast of bread and water, we became a little refreshed.

Towards noon it became fair weather; but with very little abatement of the gale, and the sea remained equally high. With great difficulty I observed the latitude to be 13° 44´ S; course N 74° W; distance 116 miles since yesterday; longitude made 31° 32´ W from Tofoa.

Sunday 24.

Sunday, May the 24th. Fresh gales and fine weather; wind S S E and S.

1789. May 24.

Towards the evening the weather looked much better, which rejoiced all hands, so that they eat their scanty allowance with more satisfaction than for some time past. The night also was fair; but, being always wet with the sea, we suffered much from the cold. A fine morning, I had the pleasure to see, produce some chearful countenances. Towards noon the weather improved, and, the first time for 15 days past, we found a little warmth from the sun. We stripped, and hung our cloaths up to dry, which were by this time become so thread-bare, that they would not keep out either wet or cold.

At noon I observed in latitude 13° 33´ S; longitude, by account, from Tofoa 33° 28´ W; course N 84° W; distance 114 miles. With the usual allowance of bread and water for dinner, I served an ounce of pork to each person.

Monday 25.

Monday, May the 25th. Fresh gales and fair weather. Wind S S E.

This afternoon we had many birds about us, which are never seen far from land, such as boobies and noddies.

Allowance lessened.

About three o'clock the sea began to run fair, and we shipped but little water, I therefore determined to know the exact quantity of bread I had left; and on examining found, according to my present issues, sufficient for 29 days allowance. In the course of this time I hoped to be at Timor; but, as that was very uncertain, and perhaps after all we might be obliged to go to Java, I determined to proportion my issues to six weeks. I was apprehensive that this would be ill received, and that it would require my utmost resolution to enforce it; for, small as the quantity was which I intended to take away, for our future good, yet it might appear to my people like robbing them of life, and some, who were less patient than their companions, I expected would very ill brook it. I however represented it so essentially necessary to guard against delays in our voyage by contrary winds, or other causes, promising to enlarge upon the allowance as we got on, that it was readily agreed to. I therefore fixed, that every person should receive one 25th of a pound of bread for breakfast, and one 25th of a pound for dinner; so that by omitting the proportion for supper, I had 43 days allowance.

1789. May 25.

At noon some noddies came so near to us, that one of them was caught by hand. This bird is about the size of a small pigeon. I divided it, with its entrails, into 18 portions, and by the method of, Who shall have this[*]? it was distributed with the allowance of bread and water for dinner, and eat up bones and all, with salt water for sauce. I observed the latitude 13° 32´ S; longitude made 35° 19´ W; and course N 89° W; distance 108 miles.

[*] One person turns his back on the object that is to be divided: another then points separately to the portions, at each of them asking aloud, "Who shall have this?" to which the first answers by naming somebody. This impartial method of division gives every man an equal chance of the best share.

Tuesday 26.

Tuesday, May the 26th. Fresh gales at S S E, and fine weather.

In the evening we saw several boobies flying so near to us, that we caught one of them by hand. This bird is as large as a good duck; like the noddy, it has received its name from seamen, for suffering itself to be caught on the masts and yards of ships. They are the most presumptive proofs of being in the neighbourhood of land of any sea-fowl we are acquainted with. I directed the bird to be killed for supper, and the blood to be given to three of the people who were the most distressed for want of food. The body, with the entrails, beak, and feet, I divided into 18 shares, and with an allowance of bread, which I made a merit of granting, we made a good supper, compared with our usual fare.

1789. May 26.

In the morning we caught another booby, so that Providence seemed to be relieving our wants in a very extraordinary manner. Towards noon we passed a great many pieces of the branches of trees, some of which appeared to have been no long time in the water. I had a good observation for the latitude, and found my situation to be in 13° 41´ S; my longitude, by account, from Tofoa, 37° 13´ W; course S 85° W, 112 miles. Every person was now overjoyed at the addition to their dinner, which I distributed as I had done in the evening; giving the blood to those who were the most in want of food.

To make our bread a little savoury we frequently dipped it in salt water; but for my own part I generally broke mine into small pieces, and eat it in my allowance of water, out of a cocoa-nut shell, with a spoon, economically avoiding to take too large a piece at a time, so that I was as long at dinner as if it had been a much more plentiful meal.

Wednesday 27.

Wednesday, May the 27th. Fresh breezes south-easterly, and fine weather.

The weather was now serene, but unhappily we found ourselves unable to bear the sun's heat; many of us suffering a languor and faintness, which made life indifferent. We were, however, so fortunate as to catch two boobies to-day; their stomachs contained several flying-fish and small cuttlefish, all of which I saved to be divided for dinner.

We passed much drift wood, and saw many birds; I therefore did not hesitate to pronounce that we were near the reefs of New Holland, and assured every one I would make the coast without delay, in the parallel we were in, and range the reef till I found an opening, through which we might get into smooth water, and pick up some supplies. From my recollection of captain Cook's survey of this coast, I considered the direction of it to be N W, and I was therefore satisfied that, with the wind to the southward of E, I could always clear any dangers.

1789. May 27.

At noon I observed in latitude 13° 26´ S; course since yesterday N 82° W; distance 109 miles; longitude made 39° 4´ W. After writing my account, I divided the two birds with their entrails, and the contents of their maws, into 18 portions, and, as the prize was a very valuable one, it was divided as before, by calling out Who shall have this? so that to-day, with the allowance of a 25th of a pound of bread at breakfast, and another at dinner, with the proportion of water, I was happy to see that every person thought he had feasted.

Thursday 28.

Thursday, May the 28th. Fresh breezes and fair weather; wind E S E and E.

In the evening we saw a gannet; and the clouds remained so fixed in the west, that I had little doubt of our being near to New Holland; and every person, after taking his allowance of water for supper, began to divert himself with conversing on the probability of what we should find.

At one in the morning the person at the helm heard the sound of breakers, and I no sooner lifted up my head, than I saw them close under our lee, not more than a quarter of a mile distant from us. I immediately hauled on a wind to the N N E, and in ten minutes time we could neither see nor hear them.

1789. May 28.

I have already mentioned my reason for making New Holland so far to the southward; for I never doubted of numerous openings in the reef, through which I could have access to the shore: and, knowing the inclination of the coast to be to the N W, and the wind mostly to the southward of E, I could with ease range such a barrier of reefs till I should find a passage, which now became absolutely necessary, without a moment's loss of time. The idea of getting into smooth water, and finding refreshments, kept my people's spirits up: their joy was very great after we had got clear of the breakers, to which we had been much nearer than I thought was possible to be before we saw them.

In the morning, at day-light, I bore away again for the reefs, and saw them by nine o'clock. The sea broke furiously over every part, and I had no sooner got near to them, than the wind came at E, so that we could only lie x0along the line of the breakers, within which we saw the water so smooth, that every person already anticipated the heart-felt satisfaction he would receive, as soon as we could get within them. But I now found we were embayed, for I could not lie clear with my sails, the wind having backed against us, and the sea set in so heavy towards the reef that our situation was become dangerous. We could effect but little with the oars, having scarce strength to pull them; and it was becoming every minute more and more probable that we should be obliged to attempt pushing over the reef, in case we could not pull off. Even this I did not despair of effecting with success, when happily we discovered a break in the reef, about one mile from us, and at the same time an island of a moderate height within it, nearly in the same direction, bearing W 1/2 N. I entered the passage with a strong stream running to the westward; and found it about a quarter of a mile broad, with every appearance of deep water.

On the outside, the reef inclined to the N E for a few miles, and from thence to the N W; on the south side of the entrance, it inclined to the S S W as far as I could see it; and I conjecture that a similar passage to this which we now entered, may be found near the breakers that I first discovered, which are 23 miles S of this channel.

1789. May 28.

I did not recollect what latitude Providential channel [*] lies in, but I considered it to be within a few miles of this, which is situate in 12° 51´ S latitude.

[*] Providential Channel is in 12° 34´ S, longitude 143° 33´ E.

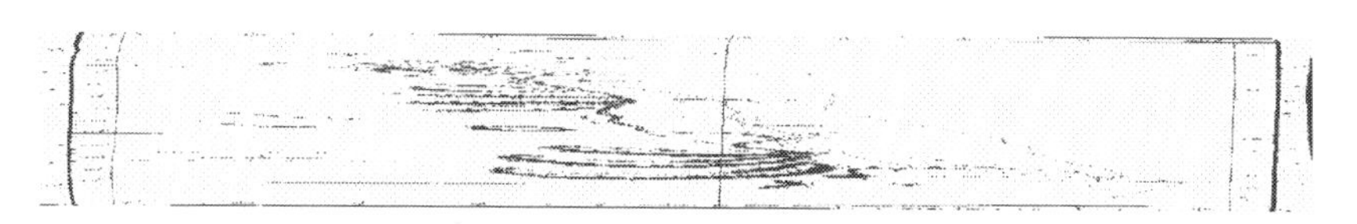

NE Coast of New Holland

Being now happily within the reefs, and in smooth water, I endeavoured to keep near them to try for fish; but the tide set us to the N W; I therefore bore away in that direction, and, having promised to land on the first convenient spot we could find, all our past hardships seemed already to be forgotten.

At noon I had a good observation, by which our latitude was 12° 46´ S, whence the foregoing situations may be considered as determined with some exactness. The island first seen bore W S W five leagues. This, which I have called the island Direction, will in fair weather always shew the channel, from which it bears due W, and may be seen as soon as the reefs, from a ship's mast-head: it lies in the latitude of 12° 51´ S. These, however, are marks too small for a ship to hit, unless it can hereafter be ascertained that passages through the reef are numerous along the coast, which I am inclined to think they are, and then there would be little risk if the wind was not directly on the shore.

My longitude, made by dead reckoning, from the island Tofoa to our passage through the reef, is 40° 10´ W. Providential channel, I imagine, must lie very nearly under the same meridian with our passage; by which it appears we had out-run our reckoning 1° 9´.

We now returned God thanks for his gracious protection, and with much content took our miserable allowance of a 25th of a pound of bread, and a quarter of a pint of water, for dinner.

Friday 29.

Friday, May the 29th. Moderate breezes and fine weather, wind E S E.

1789. May 29.

As we advanced within the reefs, the coast began to shew itself very distinctly, with a variety of high and low land; some parts of which were covered with wood. In our way towards the shore we fell in with a point of a reef, which is connected with that towards the sea, and here I came to a grapnel, and tried to catch fish, but had no success. The island Direction now bore S three or four leagues. Two islands lay about four miles to the W by N, and appeared eligible for a resting-place, if nothing more; but on my approach to the first I found it only a heap of stones, and its size too inconsiderable to shelter the boat. I therefore proceeded to the next, which was close to it and towards the main, where, on the N W side, I found a bay and a fine sandy point to land at. Our distance was about a quarter of a mile from a projecting part of the main, bearing from S W by S, to N N W 3/4 W. I now landed to examine if there were any signs of the natives being near us; but though I discovered some old fire-places, I saw nothing to alarm me for our situation during the night. Every one was anxious to find something to eat, and I soon heard that there were oysters on the rocks, for the tide was out; but it was nearly dark, and only a few could be gathered. I determined therefore to wait till the morning, to know how to proceed, and I consented that one half of us should sleep on shore, and the other in the boat. We would gladly have made a fire, but, as we could not accomplish it, we took our rest for the night, which happily was calm and undisturbed.

1789. May 29.

The dawn of day brought greater strength and spirits to us than I expected; for, notwithstanding every one was very weak, there appeared strength sufficient remaining to make me conceive the most favourable hopes of our being able to surmount the difficulties we might yet have to encounter.

As soon as I saw that there were not any natives immediately near us, I sent out parties in search of supplies, while others were putting the boat in order, that I might be ready to go to sea in case any unforeseen cause might make it necessary. The first object of this work, that demanded our attention, was the rudder: one of the gudgeons had come out, in the course of the night, and was lost. This, if it had happened at sea, would probably have been the cause of our perishing, as the management of the boat could not have been so nicely preserved as these very heavy seas required. I had often expressed my fears of this accident, and, that we might be prepared for it, had taken the precaution to have grummets fixed on each quarter of the boat for oars; but even our utmost readiness in using them, I fear, would not have saved us. It appears, therefore, a providential circumstance, that it happened at this place, and was in our power to remedy the defect; for by great good luck we found a large staple in the boat that answered the purpose.

The parties were now returned, highly rejoiced at having found plenty of oysters and fresh water. I also had made a fire, by help of a small magnifying glass, that I always carried about me, to read off the divisions of my sextants; and, what was still more fortunate, among the few things which had been thrown into the boat and saved, was a piece of brimstone and a tinder-box, so that I secured fire for the future.

1789. May 29.

One of my people had been so provident as to bring away with him a copper pot: it was by being in possession of this article that I was enabled to make a proper use of the supply we found, for, with a mixture of bread and a little pork, I made a stew that might have been relished by people of more delicate appetites, of which each person received a full pint.

The general complaints of disease among us, were a dizziness in the head, great weakness of the joints, and violent tenesmus, most of us having had no evacuation by stool since we left the ship. I had constantly a severe pain at my stomach; but none of our complaints were alarming; on the contrary, every one retained marks of strength, that, with a mind possessed of any fortitude, could bear more fatigue than I hoped we had to undergo in our voyage to Timor.

As I would not allow the people to expose themselves to the heat of the sun, it being near noon, every one took his allotment of earth, shaded by the bushes, for a short sleep.

The oysters we found grew so fast to the rocks that it was with difficulty they could be broke off, and at last we discovered it to be the most expeditious way to open them where they were found. They were very sizeable, and well tasted, and gave us great relief. To add to this happy circumstance, in the hollow of the land there grew some wire grass, which indicated a moist situation. On forcing a stick, about three feet long, into the ground, we found water, and with little trouble dug a well, which produced as much as we were in need of. It was very good, but I could not determine if it was a spring or not. Our wants made it not necessary to make the well deep, for it flowed as fast as we emptied it; which, as the soil was apparently too loose to retain water from the rains, renders it probable to be a spring. It lies about 200 yards to the S E of a point in the S W part of the island.

1789. May 29.

I found evident signs of the natives resorting to this island; for, besides fire-places, I saw two miserable wigwams, having only one side loosely covered. We found a pointed stick, about three feet long, with a slit in the end of it, to sling stones with, the same as the natives of Van Diemen's land use.

The track of some animal was very discernible, and Mr. Nelson agreed with me that it was the Kanguroo; but how these animals can get from the main I know not, unless brought over by the natives to breed, that they may take them with more ease, and render a supply of food certain to them; as on the continent the catching of them may be precarious, or attended with great trouble, in so large an extent of country.

The island may be about two miles in circuit; it is a high lump of rocks and stones covered with wood; but the trees are small, the soil, which is very indifferent and sandy, being barely sufficient to produce them. The trees that came within our knowledge were the manchineal and a species of purow: also some palm-trees, the tops of which we cut down, and the soft interior part or heart of them was so palatable that it made a good addition to our mess. Mr. Nelson discovered some fern-roots, which I thought might be good roasted, as a substitute for bread, but it proved a very poor one: it however was very good in its natural state to allay thirst, and on that account I directed a quantity to be collected to take into the boat. Many pieces of cocoa-nut shells and husk were found about the shore, but we could find no cocoa-nut trees, neither did I see any like them on the main.

1789. May 29.

I had cautioned every one not to touch any kind of berry or fruit that they might find; yet they were no sooner out of my sight than they began to make free with three different kinds, that grew all over the island, eating without any reserve. The symptoms of having eaten too much, began at last to frighten same of them; but on questioning others, who had taken a more moderate allowance, their minds were a little quieted. The others, however, became equally alarmed in their turn, dreading that such symptoms would come on, and that they were all poisoned, so that they regarded each other with the strongest marks of apprehension, uncertain what would be the issue of their imprudence. Happily the fruit proved wholesome and good. One sort grew on a small delicate kind of vine; they were the size of a large gooseberry, and very like in substance, but had only a sweet taste; the skin was a pale red, streaked with yellow the long way of the fruit: it was pleasant and agreeable. Another kind grew on bushes, like that which is called the sea-side grape in the West Indies; but the fruit was very different, and more like elder-berries, growing in clusters in the same manner. The third sort was a black berry, not in such plenty as the others, and resembled a bullace, or large kind of sloe, both in size and taste. Seeing these fruits eaten by the birds made me consider them fit for use, and those who had already tried the experiment, not finding any bad effect, made it a certainty that we might eat of them without danger.

Wild pigeons, parrots, and other birds, were about the summit of the island, but, as I had no fire-arms, relief of that kind was not to be expected, unless I met with some unfrequented spot where we might take them with our hands.

1789. May 29.

On the south side of the island, and about half a mile from the well, a small run of water was found; but, as its source was not traced, I know nothing more of it.

The shore of this island is very rocky, except the part we landed at, and here I picked up many pieces of pumice-stone. On the part of the main next to us were several sandy bays, but at low-water they became an extensive rocky flat. The country had rather a barren appearance, except in a few places where it was covered with wood. A remarkable range of rocks lay a few miles to the S W, or a high peaked hill terminated the coast towards the sea, with other high lands and islands to the southward. A high fair cape showed the direction of the coast to the N W, about seven leagues, and two small isles lay three or four leagues to the northward.

I saw a few bees or wasps, several lizards, and the blackberry bushes were full of ants nests, webbed as a spider's, but so close and compact as not to admit the rain.

A trunk of a tree, about 50 feet long, lay on the beach; from whence I conclude a heavy sea runs in here with the northerly winds.

This being the day of the restoration of king Charles the Second, and the name not being inapplicable to our present situation (for we were restored to fresh life and strength), I named this Restoration Island; for I thought it probable that captain Cook might not have taken notice of it. The other names I have presumed to give the different parts of the coast, will be only to show my route a little more distinctly.

At noon I found the latitude of the island to be 12° 39´ S; our course having been N 66° W; distance 18 miles from yesterday noon.

Saturday 30.

1789. May 30.

Saturday, May the 30th. Very fine weather, and E S E winds. This afternoon I sent parties out again to gather oysters, with which and some of the inner part of the palm-top, we made another good stew for supper, each person receiving a full pint and a half; but I refused bread to this meal, for I considered our wants might yet be very great, and as such I represented the necessity of saving our principal support whenever it was in our power.

At night we again divided, and one half of us slept on shore by a good fire. In the morning I discovered a visible alteration in every one for the better, and I sent them away again to gather oysters. I had now only two pounds of pork left. This article, which I could not keep under lock and key as I did the bread, had been pilfered by some inconsiderate person, but every one most solemnly denied it; I therefore resolved to put it out of their power for the future, by sharing what remained for our dinner. While the party was out getting oysters, I got the boat in readiness for sea, and filled all our water vessels, which amounted to nearly 60 gallons.

The party being returned, dinner was soon ready, and every one had as good an allowance as they had for supper; for with the pork I gave an allowance of bread; as I was determined forthwith to push on. As it was not yet noon, I told every one that an exertion should be made to gather as many oysters as possible for a sea store, as I was determined to sail in the afternoon.

At noon I again observed the latitude 12° 39´ S; it was then high-water, the tide had risen three feet, but I could not be certain which way the flood came from. I deduce the time of high-water at full and change to be ten minutes past seven in the morning.

Sunday 31.

1789. May 31.

Sunday, May the 31st. Early in the afternoon, the people returned with the few oysters they had time to pick up, and every thing was put into the boat. I then examined the quantity of bread remaining, and found 38 days allowance, according to the last mode of issuing a 25th of a pound at breakfast and at dinner.

Fair weather, and moderate breezes at E S E and S E.

Being all ready for sea, I directed every person to attend prayers, and by four o'clock we were preparing to embark; when twenty natives appeared, running and holloaing to us, on the opposite shore. They were armed with a spear or lance, and a short weapon which they carried in their left hand: they made signs for us to come to them. On the top of the hills we saw the heads of many more; whether these were their wives and children, or others who waited for our landing, until which they meant not to show themselves, lest we might be intimidated, I cannot say; but, as I found we were discovered to be on the coast, I thought it prudent to make the best of my way, for fear of canoes; though, from the accounts of captain Cook, the chance was that there were very few or none of any consequence. I passed these people as near as I could, which was within a quarter of a mile; they were naked, and apparently black, and their hair or wool bushy and short.

1789. May 31.

I directed my course within two small islands that lie to the north of Restoration Island, passing between them and the main land, towards Fair Cape, with a strong tide in my favour; so that I was abreast of it by eight o'clock. The coast I had passed was high and woody. As I could see no land without Fair Cape, I concluded that the coast inclined to the N W and W N W, which was agreeable to my recollection of captain Cook's survey. I therefore steered more towards the W; but by eleven o'clock at night I found myself mistaken: for we met with low land, which inclined to the N E; so that at three o'clock in the morning I found we were embayed, which obliged us to stand back to the southward.

At day-break I was exceedingly surprised to find the appearance of the country all changed, as if in the course of the night I had been transported to another part of the world; for we had now a miserable low sandy coast in view, with very little verdure, or any thing to indicate that it was at all habitable to a human being, if I except some patches of small trees or brush-wood.

1789. May 31.

I had many small islands in view to the N E, about six miles distant. The E part of the main bore N four miles, and Fair Cape S S E five or six leagues. I took the channel between the nearest island and the main land, about one mile apart, leaving all the islands on the starboard side. Some of these were very pretty spots, covered with wood, and well situated for fishing; large shoals of fish were about us, but we could not catch any. As I was passing this strait we saw another party of Indians, seven in number, running towards us, shouting and making signs for us to land. Some of them waved green branches of the bushes which were near them, as a sign of friendship; but there were some of their other motions less friendly. A larger party we saw a little farther off, and coming towards us. I therefore determined not to land, though I wished much to have had some intercourse with these people; for which purpose I beckoned to them to come near to me, and laid the boat close to the rocks; but not one would come within 200 yards of us. They were armed in the same manner as those I had seen from Restoration Island, were stark naked, and appeared to be jet black, with short bushy hair or wool, and in every respect the same people. An island of good height now bore N 1/2 W, four miles from us, at which I resolved to see what could be got, and from thence to take a look at the coast. At this isle I landed about eight o'clock in the morning. The shore was rocky, with some sandy beaches within the rocks: the water, however, was smooth, and I landed without difficulty. I sent two parties out, one to the northward, and the other to the southward, to seek for supplies, and others I ordered to stay by the boat. On this occasion their fatigue and weakness so far got the better of their sense of duty, that some of them began to mutter who had done most, and declared they would rather be without their dinner than go in search of it. One person, in particular, went so far as to tell me, with a mutinous look, he was as good a man as myself. It was not possible for me to judge where this might have an end, if not stopped in time; I therefore determined to strike a final blow at it, and either to preserve my command, or die in the attempt: and, seizing a cutlass, I ordered him to take hold of another and defend himself; on which he called out I was going to kill him, and began to make concessions. I did not allow this to interfere further with the harmony of the boat's crew, and every thing soon became quiet.

The parties continued collecting what could be found, which consisted of some fine oysters and clams, and a few small dog-fish that were caught in the holes of the rocks. We also found about two tons of rain-water in the hollow of the rocks, on the north part of the island, so that of this essential article we were again so happy as not to be in want.

1789. May 31.

After regulating the mode of proceeding, I set off for the highest part of the island, to see and consider of my route for the night. To my surprise I could see no more of the main than I did from below, it extending only from S 1/2 E, four miles, to W by N, about three leagues, full of sand-hills. Besides the isles to the E S E and south, that I had seen before, I could only discover a small key N W by N. As this was considerably farther from the main than where I was at present, I resolved to get there by night, it being a more secure resting-place; for I was here open to an attack, if the Indians had canoes, as they undoubtedly observed my landing. My mind being made up on this point, I returned, taking a particular look at the spot I was on, which I found only to produce a few bushes and coarse grass, and the extent of the whole not two miles in circuit. On the north side, in a sandy bay, I saw an old canoe, about 33 feet long, lying bottom upwards, and half buried in the beach. It was made of three pieces, the bottom entire, to which the sides were sewed in the common way. It had a sharp projecting prow rudely carved, in resemblance of the head of a fish; the extreme breadth was about three feet, and I imagine it was capable of carrying 20 men.

At noon the parties were all returned, but had found difficulty in gathering the oysters, from their close adherence to the rocks, and the clams were scarce: I therefore saw, that it would be of little use to remain longer in this place, as we should not be able to collect more than we could eat; nor could any tolerable sea-store be expected, unless we fell in with a greater plenty. I named this Sunday Island: it lies N by W 3/4 W from Restoration Island; the latitude, by a good observation, 11° 58′ S.

JUNE

June. Monday 1.

Monday, June the 1st. Fresh breezes and fair weather, ending with a fresh gale. Wind S E by S.

1789. June 1.

At two o'clock in the afternoon, we dined; each person having a full pint and a half of stewed oysters and clams, thickened with small beans, which Mr. Nelson informed us were a species of Dolichos. Having eaten heartily, and taken the water we were in want of, I only waited to determine the time of high-water, which I found to be at three o'clock, and the rise of the tide about five feet. According to this it is high-water on the full and change at 19 minutes past 9 in the morning; but here I observed the flood to come from the southward, though at Restoration Island, I thought it came from the northward. I think captain Cook mentions that he found great irregularity in the set of the flood on this coast.

I now sailed for the key which I had seen in the N W by N, giving the name of Sunday Island to the place I left; we arrived just at dark, but found it so surrounded by a reef of rocks, that I could not land without danger of staving the boat; and on that account I came to a grapnel for the night.

1789. June 1.

At dawn of day we got on shore, and tracked the boat into shelter; for the wind blowing fresh without, and the ground being rocky, I was afraid to trust her at a grapnel, lest she might be blown to sea: I was, therefore, obliged to let her ground in the course of the ebb. From appearances, I expected that if we remained till night we should meet with turtle, as we had already discovered recent tracks of them. Innumerable birds of the noddy kind made this island their resting-place; so that I had reason to flatter myself with hopes of getting supplies in greater abundance than it had hitherto been in my power. The situation was at least four leagues distant from the main. We were on the north-westernmost of four small keys, which were surrounded by a reef of rocks connected by sand-banks, except between the two northernmost; and there likewise it was dry at low water; the whole forming a lagoon island, into which the tide flowed: at this entrance I kept the boat.

As usual, I sent parties away in search of supplies, but, to our great disappointment, we could only get a few clams and some dolichos: with these, and the oysters we had brought from Sunday Island, I made up a mess for dinner, with an addition of a small quantity of bread.

1789. June 1.

Towards noon, Mr. Nelson, and his party, who had been to the easternmost key, returned; but himself in such a weak condition, that he was obliged to be supported by two men. His complaint was a violent heat in his bowels, a loss of sight, much drought, and an inability to walk. This I found was occasioned by his being unable to support the heat of the sun, and that, when he was fatigued and faint, instead of retiring into the shade to rest, he had continued to do more than his strength was equal to. It was a great satisfaction to me to find, that he had no fever; and it was now that the little wine, which I had so carefully saved became of real use. I gave it in very small quantities, with some small pieces of bread soaked in it; and, having pulled off his cloaths, and laid him under some shady bushes, he began to recover. The boatswain and carpenter also were ill, and complained of head-ach, and sickness of the stomach; others, who had not had any evacuation by stool, became shockingly distressed with the tenesmus; so that there were but few without complaints. An idea now prevailed, that their illness was occasioned by eating the dolichos, and some were so much alarmed that they thought themselves poisoned. Myself, however, and some others, who had eaten of them, were yet very well; but the truth was, that all those who were complaining, except Mr. Nelson, had gorged themselves with a large quantity of raw beans, and Mr. Nelson informed me, that they were constantly teazing him, whenever a berry was found, to know if it was good to eat; so that it would not have been surprizing if many of them had been really poisoned.

Our dinner was not so well relished as at Sunday Island, because we had mixed the dolichos with our stew. The oysters and soup, however, were eaten by every one, except Mr. Nelson, whom I fed with a few small pieces of bread soaked in half a glass of wine, and he continued to mend.

In my walk round the island, I found several cocoa-nut shells, the remains of an old wigwam, and the backs of two turtle, but no sign of any quadruped. One of my people found three sea-fowl's eggs.

As is common on such spots, the soil is little other than sand, yet it produced small toa-trees, and some others, that we were not acquainted with. There were fish in the lagoon, but we could not catch any. As our wants, therefore, were not likely to be supplied here, not even with water for our daily expence, I determined to sail in the morning, after trying our success in the night for turtle and birds. A quiet night's rest also, I conceived, would be of essential service to those who were unwell.

From the wigwam and turtle-shell being found, it is certain that the natives sometimes resort to this place, and have canoes: but I did not apprehend that we ran any risk by remaining here. I directed our fire, however, to be made in the thicket, that we might not be discovered in the night.

1789. June 1.

At noon, I observed the latitude of this island to be 11° 47′ S. The main land extended towards the N W, and was full of white sand-hills:

another small island lay within us, bearing W by N 1/4 N, three leagues distant. My situation being very low, I could see nothing of the reef towards the sea.

Tuesday 2.

Tuesday, June the 2d. The first part of this day we had some light showers of rain; the latter part was fair, wind from the S E, blowing fresh.

Rest was now so much wanted, that the afternoon was advantageously spent in sleep. There were, however, a few not disposed to it, and those I employed in dressing some clams to take with us for the next day's dinner; others we cut up in slices to dry, which I knew was the most valuable supply we could find here. But, contrary to our expectation, they were very scarce.

1789. June 2.

Towards evening, I cautioned every one against making too large a fire, or suffering it after dark to blaze up. Mr. Samuel and Mr. Peckover had the superintendence of this business, while I was strolling about the beach to observe if I thought it could be seen from the main. I was just satisfied that it could not, when on a sudden the island appeared all in a blaze, that might have been seen at a much more considerable distance. I ran to learn the cause, and found it was occasioned by the imprudence and obstinacy of one of the party, who, in my absence, had insisted on having a fire to himself; in making which the flames caught the neighbouring grass and rapidly spread. This misconduct might have produced very serious consequences, by discovering our situation to the natives; for, if they had attacked us, we must inevitably have fallen a sacrifice, as we had neither arms nor strength to oppose an enemy. Thus the relief which I expected from a little sleep was totally lost, and I anxiously waited for the flowing of the tide, that we might proceed to sea.

I found it high-water at half past five this evening, whence I deduce the time, on the full and change of the moon, to be 58' past 10 in the morning: the rise is nearly five feet. I could not observe the set of the flood; but imagine it comes from the southward, and that I have been mistaken at Restoration Island, as I find the time of high-water gradually later as we advance to the northward.

At Restoration Island, high water, full and change,	7^h 10'
Sunday Island,	9 19
Here,	10 58

1789. June 2.

After eight o'clock, Mr. Samuel and Mr. Peckover went out to watch for turtle, and three men went to the east key to endeavour to catch birds. All the others complaining of being sick, took their rest, except Mr. Hayward and Mr. Elphinston, who I directed to keep watch. About midnight the bird party returned, with only twelve noddies, a bird I have already described to be about the size of a pigeon: but if it had not been for the folly and obstinacy of one of the party, who separated from the other two, and disturbed the birds, they might have caught a great number. I was so much provoked at my plans being thus defeated, that I gave the offender[*] a good beating. I now went in search of the turtling party, who had taken great pains, but without success. This, however, did not surprise me, as it was not to be expected that turtle would come near us after the noise which was made at the beginning of the evening in extinguishing the fire. I therefore desired them to come back, but they requested to stay a little longer, as they still hoped to find some before day-light: they, however, returned by three o'clock, without any reward for their labour.

[*] Robert Lamb.—This man, when he came to Java, acknowledged he had eaten nine birds on the key, after he separated from the other two.

The birds we half dressed, which, with a few clams, made the whole of the supply procured here. I tied up a few gilt buttons and some pieces of iron to a tree, for any of the natives that might come after us; and, happily finding my invalids much better for their night's rest, I got every one into the boat, and departed by dawn of day. Wind at S E; course to the N by W.

We had scarcely ran two leagues to the northward, when the sea suddenly became rough, which not having experienced since we were within the reefs, I concluded to be occasioned by an open channel to the ocean. Soon afterwards we met with a large shoal, on which were two sandy keys; between these and two others, four miles to the west, I passed on to the northward, the sea still continuing to be rough.

1789. June 2.

Towards noon, I fell in with six other keys, most of which produced some small trees and brush-wood. These formed a pleasing contrast with the main land we had passed, which was full of sand-hills. The country continued hilly, and the northernmost land, the same which we saw from the lagoon island, appeared like downs, sloping towards the sea. To the southward of this is a flat-topped hill, which, on account of its shape, I called Pudding-pan hill, and a little to the northward two other hills, which we called the Paps; and here was a small tract of country without sand, the eastern part of which forms a cape, whence the coast inclines to the N W by N.

At noon I observed in the latitude of 11° 18´ S, the cape bearing W, distant ten miles. Five small keys bore from N E to S E, the nearest of them about two miles distant, and a low sandy key between us and the cape bore W, distant four miles. My course from the Lagoon Island N 1/2 W, distant 30 miles.

I am sorry it was not in my power to obtain a sufficient knowledge of the depth of water; for in our situation nothing could be undertaken that might have occasioned delay. It may however be understood, that, to the best of my judgment, from appearances, a ship may pass wherever I have omitted to represent danger.

I divided six birds, and issued one 25th of a pound of bread, with half a pint of water, to each person for dinner, and I gave half a glass of wine to Mr. Nelson, who was now so far recovered as to require no other indulgence.

The gunner, when he left the ship, brought his watch with him, by which we had regulated our time till to-day, when unfortunately it stopped; so that noon, sun-rise, and sun-set, are the only parts of the 24 hours of which I can speak with certainty, as to time.

Wednesday 3.

1789. June 3.

1789 June 3.

Wednesday, June the 3d. Fresh gales S S E and S E, and fair weather. As we stood to the N by W this afternoon, we found more sea, which I attributed to our receiving less shelter from the reefs to the eastward: it is probable they do not extend so far to the N as this; at least, it may be concluded that there is not a continued barrier to prevent shipping having access to the shore. I observed that the stream set to the N W, which I considered to be the flood; in some places along the coast, we saw patches of wood. At five o'clock, steering to the N W, we passed a large and fair inlet, into which, I imagine, is a safe and commodious entrance; it lies in latitude 11° S: about three leagues to the northward of this is an island, at which we arrived about sun-set, and took shelter for the night under a sandy point, which was the only part we could land at: I was therefore under the necessity to put up with rather a wild situation, and slept in the boat. Nevertheless I sent a party away to see what could be got, but they returned without any success. They saw a great number of turtle bones and shells, where the natives had been feasting, and their last visit seemed to be of late date. The island was covered with wood, but in other respects a lump of rocks. We lay at a grapnel until day-light, with a very fresh gale and cloudy weather. The main bore from S E by S to N N W 1/2 W, three leagues; and a mountainous island, with a flat top, N by W, four or five leagues: several others were between it and the main. The spot we were on, which I call Turtle Island; lies in latitude, by account, 10° 52′ S, and 42 miles W from Restoration Island. Abreast of it the coast has the appearance of a sandy desert, but improves about three leagues farther to the northward, where it terminates in a point, near to which is a number of small islands. I sailed between these islands, where I found no bottom at twelve fathoms; the high mountainous island with a flat top, and four rocks to the S E of it, that I call the Brothers, being on my starboard hand. Soon after, an extensive opening appeared in the main land, with a number of high islands in it. I called this the Bay of Islands. We continued steering to the N W. Several islands and keys lay to the northward. The most northerly island was mountainous, having on it a very high round hill; and a smaller was remarkable for a single peaked hill.

The coast to the northward and westward of the Bay of Islands had a very different appearance from that to the southward. It was high and woody, with many islands close to it, and had a very broken appearance. Among these islands are fine bays, and convenient places for shipping. The northernmost I call Wednesday Island: to the N W of this we fell in with a large reef, which I believe joins a number of keys that were in sight from the N W to the E N E. We now stood to the S W half a league, when it was noon, and I had a good observation of the latitude in 10° 31´ S. Wednesday Island bore E by S five miles; the westernmost land S W two or three leagues; the islands to the northward, from N W by W four or five leagues, to N E six leagues; and the reef from W to N E, distant one mile, I now assured every one that we should be clear of New Holland in the afternoon.

It is impossible for me to say how far this reef may extend. It may be a continuation, or a detached part of the range of shoals that surround the coast: but be that as it may, I consider the mountainous islands as separate from the shoals; and have no doubt that near them may be found good passages for ships. But I rather recommend to those who are to pass this strait from the eastward, to take their direction from the coast of New Guinea: yet, I likewise think that a ship coming from the southward, will find a fair strait in the latitude of 10° S. I much wished to have ascertained this point; but in our distressful situation, any increase of fatigue, or loss of time, might have been attended with the most fatal consequences. I therefore determined to pass on without delay.

1789. June 3.

As an addition to our dinner of bread and water, I served to each person six oysters.

Thursday 4.

Thursday, June the 4th. A fresh gale at S E, and fair weather.

At two o'clock as we were steering to the S W, towards the westernmost part of the land in sight, we fell in with some large sand-banks that run off from the coast. We were therefore obliged to steer to the northward again, and, having got round them, I directed my course to the W.

At four o'clock, the westernmost of the islands to the northward bore N four leagues; Wednesday island E by N five leagues; and Shoal Cape S E by E two leagues. A small island was now seen bearing W, at which I arrived before dark, and found that it was only a rock, where boobies resort, for which reason I called it Booby Island. A small key also lies close to the W part of the coast, which I have called Shoal Cape. Here terminated the rocks and shoals of the N part of New Holland, for, except Booby Island, we could see no land to the westward of S, after three o'clock this afternoon.

1789. June 4.

I find that Booby Island was seen by Captain Cook, and, by a remarkable coincidence of ideas, received from him the same name; but I cannot with certainty reconcile the situation of many parts of the coast that I have seen, to his survey. I ascribe this to the very different form in which land appears, when seen from the unequal heights of a ship and a boat. The chart I have given, is by no means meant to supersede that made by Captain Cook, who had better opportunities than I had, and was in every respect properly provided for surveying. The intention of mine is chiefly to render the narrative more intelligible, and to shew in what manner the coast appeared to me from an open boat. I have little doubt that the opening, which I named the Bay of Islands, is Endeavour Straits; and that our track was to the northward of Prince of Wales's Isles. Perhaps, by those who shall hereafter navigate these seas, more advantage may be derived from the possession of both our charts, than from either singly.

At eight o'clock in the evening, we once more launched into the open ocean. Miserable as our situation was in every respect, I was secretly surprised to see that it did not appear to affect any one so strongly as myself; on the contrary, it seemed as if they had embarked on a voyage to Timor, in a vessel sufficiently calculated for safety and convenience.

So much confidence gave me great pleasure, and I may assert that to this cause their preservation is chiefly to be attributed; for if any one of them had despaired, he would most probably have died before we reached New Holland.

I now gave every one hopes that eight or ten days might bring us to a land of safety; and, after praying to God for a continuance of his most gracious protection, I served an allowance of water for supper, and kept my course to the W S W, to counteract the southerly winds, in case they should blow strong.

1789. June 4.

We had been just six days on the coast of New Holland, in the course of which we found oysters, a few clams, some birds, and water. But perhaps a benefit nearly equal to this we received from not having fatigue in the boat, and enjoying good rest at night. These advantages certainly preserved our lives; for, small as the supply was, I am very sensible how much it relieved our distresses. About this time nature would have sunk under the extremes of hunger and fatigue. Some would have ceased to struggle for a life that only promised wretchedness and misery; while others, though possessed of more bodily strength, must soon have followed their unfortunate companions. Even in our present situation, we were most wretched spectacles; yet our fortitude and spirit remained; every one being encouraged by the hopes of a speedy termination to his misery.

For my own part, wonderful as it may appear, I felt neither extreme hunger nor thirst. My allowance contented me, knowing I could have no more.

I served one 25th of a pound of bread, and an allowance of water, for breakfast, and the same for dinner, with an addition of six oysters to each person. At noon, latitude observed 10° 48´ S; course since yesterday noon S 81 W; distance 111 miles; longitude, by account, from Shoal Cape 1° 45´ W.

Friday 5.

Friday, June the 5th. Fair weather with some showers, and a strong trade wind at E S E.

This day we saw a number of water-snakes, that were ringed yellow and black, and towards noon we passed a great deal of rock-weed. Though the weather was fair, we were constantly shipping water, and two men always employed to bale the boat.

At noon I observed in latitude 10° 45′ S; our course since yesterday W 1/4 N, 108 miles; longitude made 3° 35′ W. Served one 25th of a pound of bread, and a quarter of a pint of water for breakfast; the same for dinner, with an addition of six oysters; for supper water only.

Saturday 6.

Saturday, June the 6th. Fair weather, with some showers, and a fresh gale at S E and E S E. Constantly shipping water and baling.

1789. June 6.

In the evening a few boobies came about us, one of which I caught with my hand. The blood was divided among three of the men who were weakest, but the bird I ordered to be kept for our dinner the next day. Served a quarter of a pint of water for supper, and to some, who were most in need, half a pint.

In the course of the night we suffered much cold and shiverings. At day-light, I found that some of the clams, which had been hung up to dry for sea-store, were stolen; but every one most solemnly denied having any knowledge of it. This forenoon we saw a gannet, a sand-lark, and some water-snakes, which in general were from two to three feet long.

Served the usual allowance of bread and water for breakfast, and the same for dinner, with the bird, which I distributed in the usual way, of Who shall have this? I determined to make Timor about the latitude of 9° 30′ S, or 10° S. At noon I observed the latitude to be 10° 19′ S; course N 77° W; distance 117 miles; longitude made from the Shoal Cape, the north part of New Holland, 5° 31′ W.

Sunday 7.

Sunday, June the 7th. Fresh gales and fair weather till eight in the evening. The remaining part of the 24 hours squally, with much wind at S S E and E S E, and a high sea, so that we were constantly wet and baling.

In the afternoon, I took an opportunity of examining again into our store of bread, and found remaining 19 days allowance, at my former rate of serving one 25th of a pound three times a day: therefore, as I saw every prospect of a quick passage, I again ventured to grant an allowance for supper, agreeable to my promise at the time it was discontinued.

1789. June 7.

We passed the night miserably wet and cold, and in the morning I heard heavy complaints of our deplorable situation. The sea was high and breaking over us. I could only afford the allowance of bread and water for breakfast; but for dinner I gave out an ounce of dried clams to each person, which was all that remained.

At noon I altered the course to the W N W, to keep more from the sea while it blew so strong. Latitude observed 9° 31´ S; course N 57° W; distance 88 miles; longitude made 6° 46´ W.

Monday 8.

Monday, June the 8th. Fresh gales and squally weather, with some showers of rain. Wind E and E S E.

This day the sea ran very high, and we were continually wet, suffering much cold in the night. I now remarked that Mr. Ledward, the surgeon, and Lawrence Lebogue, an old hardy seaman, were giving way very fast. I could only assist them by a tea-spoonful or two of wine, which I had carefully saved, expecting such a melancholy necessity. Among most of the others I observed more than a common inclination to sleep, which seemed to indicate that nature was almost exhausted.

Served the usual allowance of bread and water at supper, breakfast, and dinner. Saw several gannets.

At noon I observed in 8° 45´ S; course W N W 1/4 W, 106 miles; longitude made 8° 23´ W.

Tuesday 9.

Tuesday, June the 9th. Wind S E. The weather being moderate, I steered W by S.

1789 June 9.

At four in the afternoon we caught a small dolphin, the first relief of the kind we obtained. I issued about two ounces to each person, including the offals, and saved the remainder for dinner the next day. Towards evening the wind freshened, and it blew strong all night, so that we shipped much water, and suffered greatly from the wet and cold. At day-light, as usual, I heard much complaining, which my own feelings convinced me was too well founded. I gave the surgeon and Lebogue a little wine, but I could give no farther relief, than assurances that a very few days longer, at our present fine rate of sailing, would bring us to Timor.

Gannets, boobies, men of war and tropic birds, were constantly about us. Served the usual allowance of bread and water, and at noon dined on the remains of the dolphin, which amounted to about an ounce per man. I observed the latitude to be 9° 9´ S; longitude made 10° 8´ W; course since yesterday noon S 76° W; distance 107 miles.

Wednesday 10.

Wednesday, June the 10th. Wind E S E. Fresh gales and fair weather, but a continuance of much sea, which, by breaking almost constantly over the boat, made us miserably wet, and we had much cold to endure in the night.

This afternoon I suffered great sickness from the oily nature of part of the stomach of the fish, which had fallen to my share at dinner. At sun-set I served an allowance of bread and water for supper. In the morning, after a very bad night, I could see an alteration for the worse in more than half my people. The usual allowance was served for breakfast and dinner. At noon I found our situation to be in latitude 9° 16´ S; longitude from the north part of New Holland 12° 1´ W; course since yesterday noon W 1/2 S, distance 111 miles.

Thursday 11.

Thursday, June the 11th. Fresh gales and fair weather. Wind S E and S S E.

1789. June 11.

Birds and rock-weed showed that we were not far from land; but I expected such signs must be here, as there are many islands between the east part of Timor and New Guinea. I however hoped to fall in with Timor every hour, for I had great apprehensions that some of my people could not hold out. An extreme weakness, swelled legs, hollow and ghastly countenances, great propensity to sleep, with an apparent debility of understanding, seemed to me melancholy presages of their approaching dissolution. The surgeon and Lebogue, in particular were most miserable objects. I occasionally gave them a few tea-spoonfuls of wine, out of the little I had saved for this dreadful stage, which no doubt greatly helped to support them.

For my own part, a great share of spirits, with the hopes of being able to accomplish the voyage, seemed to be my principal support; but the boatswain very innocently told me, that he really thought I looked worse than any one in the boat. The simplicity with which he uttered such an opinion diverted me, and I had good humour enough to return him a better compliment.

Every one received his 25th of a pound of bread, and quarter of a pint of water, at evening, morning, and noon, and an extra allowance of water was given to those who desired it.

At noon I observed in latitude 9° 41´ S; course S 77° W; distance 109 miles; longitude made 13° 49´ W. I had little doubt of having now passed the meridian of the eastern part of Timor, which is laid down in 128° E. This diffused universal joy and satisfaction.

Friday 12.

Friday, June the 12th. Fresh breezes and fine weather, but very hazy. Wind from E to S E.

All the afternoon we had several gannets, and many other birds, about us, that indicated we were near land, and at sun-set we kept a very anxious look-out. In the evening we caught a booby, which I reserved for our dinner the next day.

1789. June 12.

At three in the morning, with an excess of joy, we discovered Timor bearing from W S W to W N W, and I hauled on a wind to the N N E till day-light, when the land bore from S W by S about two leagues to N E by N seven leagues.

It is not possible for me to describe the pleasure which the blessing of the sight of land diffused among us. It appeared scarce credible, that in an open boat, and so poorly provided, we should have been able to reach the coast of Timor in forty-one days after leaving Tofoa, having in that time run, by our log, a distance of 3618 miles, and that, notwithstanding our extreme distress, no one should have perished in the voyage.

I have already mentioned, that I knew not where the Dutch settlement was situated; but I had a faint idea that it was at the S W part of the island. I therefore, after day-light, bore away along shore to the S S W, and the more readily as the wind would not suffer us to go towards the N E without great loss of time.

1789. June 12.

The day gave us a most agreeable prospect of the land, which was interspersed with woods and lawns; the interior part mountainous, but the shore low. Towards noon the coast became higher, with some remarkable head-lands. We were greatly delighted with the general look of the country, which exhibited many cultivated spots and beautiful situations; but we could only see a few small huts, whence I concluded no European resided in this part of the island. Much sea ran on the shore, so that landing with a boat was impracticable. At noon I was abreast of a very high head-land; the extremes of the land bore S W 1/2 W, and N N E 1/2 E; our distance off shore being three miles; latitude, by observation, 9° 59´ S; and my longitude, by dead reckoning, from the north part of New Holland, 15° 6´ W.

With the usual allowance of bread and water for dinner, I divided the bird we had caught the night before, and to the surgeon and Lebogue I gave a little wine.

Saturday 13.

Saturday, June the 13th. Fresh gales at E, and E S E, with very hazy weather.

During the afternoon, we continued our course along a low woody shore, with innumerable palm-trees, called the Fan Palm from the leaf spreading like a fan; but we had now lost all signs of cultivation, and the country had not so fine an appearance as it had to the eastward. This, however, was only a small tract, for by sun-set it improved again, and I saw several great smokes where the inhabitants were clearing and cultivating their grounds. We had now ran 25 miles to the W S W since noon, and were W five miles from a low point, which in the afternoon I imagined had been the southernmost land, and here the coast formed a deep bend, with low land in the bight that appeared like islands. The west shore was high; but from this part of the coast to the high cape which we were abreast of yesterday noon, the shore is low, and I believe shoal. I particularly remark this situation, because here the very high ridge of mountains, that run from the east end of the island, terminate, and the appearance of the country suddenly changes for the worse, as if it was not the same island in any respect.

1789. June 13.

That we might not run past any settlement in the night, I determined to preserve my station till the morning, and therefore hove to under a close-reefed fore-sail, with which the boat lay very quiet. We were here in shoal water; our distance from the shore being half a league, the westernmost land in sight bearing W S W 1/2 W. Served bread and water for supper, and the boat lying too very well, all but the officer of the watch endeavoured to get a little sleep.

At two in the morning, we wore, and stood in shore till day-light, when I found we had drifted, during the night, about three leagues to the W S W, the southernmost land in sight bearing W. On examining the coast, and not seeing any sign of a settlement, we bore away to the westward, having a strong gale, against a weather current, which occasioned much sea. The shore was high and covered with wood, but we did not run far before low land again formed the coast, the points of which opening at west, I once more fancied we were on the south part of the island; but at ten o'clock we found the coast again inclining towards the south, part of it bearing W S W 1/2 W. At the same time high land appeared from S W to S W by W 1/2 W; but the weather was so hazy, that it was doubtful whether the two lands were separated, the opening only extending one point of the compass. I, for this reason, stood towards the outer land, and found it to be the island Roti.

1789. June 13.

I returned to the shore I had left, and in a sandy bay I brought to a grapnel, that I might more conveniently calculate my situation. In this place we saw several smokes, where the natives were clearing their grounds. During the little time we remained here, the master and carpenter very much importuned me to let them go in search of supplies; to which, at length, I assented; but, finding no one willing to be of their party, they did not choose to quit the boat. I stopped here no longer than for the purpose just mentioned, and we continued steering along shore. We had a view of a beautiful-looking country, as if formed by art into lawns and parks. The coast is low, and covered with woods, in which are innumerable fan palm-trees, that look like cocoa-nut walks. The interior part is high land, but very different from the more eastern parts of the island, where it is exceedingly mountainous, and to appearance the soil better.

At noon, the island Roti bore S W by W seven leagues. I had no observation for the latitude, but, by account, we were in 10° 12´ S; our course since yesterday noon being S 77 W, 54 miles. The usual allowance of bread and water was served for breakfast and dinner, and to the surgeon and Lebogue, I gave a little wine.

Sunday 14.

Sunday, June the 14th. A strong gale at E S E, with hazy weather, all the afternoon; after which the wind became moderate.

At two o'clock this afternoon, having run through a very dangerous breaking sea, the cause of which I attributed to a strong tide setting to windward, and shoal water, we discovered a spacious bay or sound, with a fair entrance about two or three miles wide. I now conceived hopes that our voyage was nearly at an end, as no place could appear more eligible for shipping, or more likely to be chosen for an European settlement: I therefore came to a grapnel near the east side of the entrance, in a small sandy bay, where we saw a hut, a dog, and some cattle; and I immediately sent the boatswain and gunner away to the hut, to discover the inhabitants.

The S W point of the entrance bore W 1/2 S three miles; the S E point S by W three quarters of a mile; and the island Roti from S by W 1/4 W to S W 1/4 W, about five leagues.

1789. June 14.

While we lay here I found the ebb came from the northward, and before our departure the falling of the tide discovered to us a reef of rocks, about two cables length from the shore; the whole being covered at high-water, renders it dangerous. On the opposite shore also appeared very high breakers; but there is nevertheless plenty of room, and certainly a safe channel for a first-rate man of war.

The bay or sound within, seemed to be of a considerable extent; the northern part, which I had now in view, being about five leagues distant. Here the land made in moderate risings joined by lower grounds. But the island Roti, which lies to the southward, is the best mark to know this place.

I had just time to make these remarks, when I saw the boatswain and gunner returning with some of the natives. I therefore no longer doubted of our success, and that our most sanguine expectations would be fully gratified. They brought five Indians, and informed me that they had found two families, where the women treated them with European politeness. From these people I learned, that the governor resided at a place called Coupang, which was some distance to the N E. I made signs for one of them to go in the boat, and show me Coupang, intimating that I would pay him for his trouble; the man readily complied, and came into the boat.

1789. June 14.

These people were of a dark tawny colour, and had long black hair; they chewed a great deal of beetle, and wore a square piece of cloth round their hips, in the folds of which was stuck a large knife. They had a handkerchief wrapped round their heads, and at their shoulders hung another tied by the four corners, which served as a bag for their beetle equipage.

They brought us a few pieces of dried turtle, and some ears of Indian corn. This last was most welcome to us; for the turtle was so hard, that it could not be eaten without being first soaked in hot water. Had I staid they would have brought us something more; but, as the pilot was willing, I was determined to push on. It was about half an hour past four when we sailed.

By direction of the pilot we kept close to the east shore under all our sail; but as night came on, the wind died away, and we were obliged to try at the oars, which I was surprised to see we could use with some effect. However, at ten o'clock, as I found we got but little ahead, I came to a grapnel, and for the first time I issued double allowance of bread and a little wine to each person.

At one o'clock in the morning, after the most happy and sweet sleep that ever men had, we weighed, and continued to keep the east shore on board, in very smooth water; when at last I found we were again open to the sea, the whole of the land to the westward, that we had passed, being an island, which the pilot called Pulo Samow. The northern entrance of this channel is about a mile and a half or two miles wide, and I had no ground at ten fathoms.

1789. June 14.

Hearing the report of two cannon that were fired, gave new life to every one; and soon after we discovered two square-rigged vessels and a cutter at anchor to the eastward. I endeavoured to work to windward, but we were obliged to take to our oars again, having lost ground on each tack. We kept close to the shore, and continued rowing till four o'clock, when I brought to a grapnel, and gave another allowance of bread and wine to all hands. As soon as we had rested a little, we weighed again, and rowed till near day-light, when I came to a grapnel, off a small fort and town, which the pilot told me was Coupang.

Among the things which the boatswain had thrown into the boat before we left the ship, was a bundle of signal flags that had been made for the boats to show the depth of water in sounding; with these I had, in the course of the passage, made a small jack, which I now hoisted in the main shrouds, as a signal of distress; for I did not choose to land without leave.

Soon after day-break a soldier hailed me to land, which I instantly did, among a croud of Indians, and was agreeably surprised to meet with an English sailor, who belonged to one of the vessels in the road. His captain, he told me, was the second person in the town; I therefore desired to be conducted to him, as I was informed the governor was ill, and could not then be spoken with.

Captain Spikerman received me with great humanity. I informed him of our miserable situation; and requested that care might be taken of those who were with me, without delay. On which he gave directions for their immediate reception at his own house, and went himself to the governor, to know at what time I could be permitted to see him; which was fixed to be at eleven o'clock.

I now desired every one to come on shore, which was as much as some of them could do, being scarce able to walk: they, however, got at last to the house, and found tea with bread and butter provided for their breakfast.

1789. June 14.

The abilities of a painter, perhaps, could never have been displayed to more advantage than in the delineation of the two groups of figures, which at this time presented themselves. An indifferent spectator would have been at a loss which most to admire; the eyes of famine sparkling at immediate relief, or the horror of their preservers at the sight of so many spectres, whose ghastly countenances, if the cause had been unknown, would rather have excited terror than pity. Our bodies were nothing but skin and bones, our limbs were full of sores, and we were cloathed in rags; in this condition, with the tears of joy and gratitude flowing down our cheeks, the people of Timor beheld us with a mixture of horror, surprise, and pity.

1789. June 14.

The governor, Mr. William Adrian Van Este, notwithstanding his extreme ill-health, became so anxious about us, that I saw him before the appointed time. He received me with great affection, and gave me the fullest proofs that he was possessed of every feeling of a humane and good man. Sorry as he was, he said, that such a calamity could ever have happened to us, yet he considered it as the greatest blessing of his life that we had fallen under his protection; and, though his infirmity was so great that he could not do the office of a friend himself, he would give such orders as I might be certain would procure me every supply I wanted. In the mean time a house was hired for me, and, till matters could be properly regulated, victuals for every one were ordered to be dressed at his own house. With respect to my people, he said I might have room for them either at the hospital or on board of captain Spikerman's ship, which lay in the road; and he expressed much uneasiness that Coupang could not afford them better accommodations, the house assigned to me being the only one uninhabited, and the situation of the few families such, that they could not accommodate any one. After this conversation an elegant repast was set before me, more according to the custom of the country, than with design to alleviate my hunger: so that in this instance he happily blended, with common politeness, the greatest favour I could receive.

On returning to my people, I found every kind relief had been given to them. The surgeon had dressed their sores, and the cleaning of their persons had not been less attended to, besides several friendly gifts of apparel.

I now desired to be shewn to the house that was intended for me, and I found it ready, with servants to attend, and a particular one, which the governor had directed to be always about my person. The house consisted of a hall, with a room at each end, and a loft over-head; and was surrounded by a piazza, with an outer apartment in one corner, and a communication from the back part of the house to the street. I therefore determined, instead of separating from my people, to lodge them all with me; and I divided the house as follows: One room I took to myself, the other I allotted to the master, surgeon, Mr. Nelson, and the gunner; the loft to the other officers; and the outer apartment to the men. The hall was common to the officers, and the men had the back piazza. Of this I informed the governor, and he sent down chairs, tables, and benches, with bedding and other necessaries for the use of every one.

1789. June 14.

The governor, when I took my leave, had desired me to acquaint him with every thing of which I stood in need; but I was now informed it was only at particular times that he had a few moments of ease, or could attend to any thing; being in a dying state, with an incurable disease. On this account, whatever business I had to transact would be with Mr. Timotheus Wanjon, the second of this place, and the governor's son-in-law; who now also was contributing every thing in his power to make our situation comfortable. I had been, therefore, misinformed by the seaman, who told me that captain Spikerman was the next person to the governor.

At noon a very handsome dinner was brought to the house, which was sufficient to make persons, more accustomed to plenty, eat too much.

Cautions, therefore, might be supposed to have had little effect; but I believe few people in such a situation would have observed more moderation. My greatest apprehension was, that they would eat too much fruit.

Having seen every one enjoy this meal of plenty, I dined with Mr. Wanjon; but I found no extraordinary inclination to eat or drink. Rest and quiet, I considered, as more necessary to my doing well, and therefore retired to my room, which I found furnished with every convenience. But, instead of rest, my mind was disposed to reflect on our late sufferings, and on the failure of the expedition; but, above all, on the thanks due to Almighty God, who had given us power to support and bear such heavy calamities, and had enabled me at last to be the means of saving eighteen lives.

1789. June 14.

In times of difficulty there will generally arise circumstances that bear more particularly hard on a commander. In our late situation, it was not the least of my distresses, to be constantly assailed with the melancholy demands of my people for an increase of allowance, which it grieved me to refuse. The necessity of observing the most rigid œconomy in the distribution of our provisions was so evident, that I resisted their solicitations, and never deviated from the agreement we made at setting out. The consequence of this care was, that at our arrival we had still remaining sufficient for eleven days, at our scanty allowance: and if we had been so unfortunate as to have missed the Dutch settlement at Timor, we could have proceeded to Java, where I was certain every supply we wanted could be procured.

Another disagreeable circumstance, to which my situation exposed me, was the caprice of ignorant people. Had I been incapable of acting, they would have carried the boat on shore as soon as we made the island of Timor, without considering that landing among the natives, at a distance from the European settlement, might have been as dangerous as among any other Indians.

The quantity of provisions with which we left the ship, was not more than we should have consumed in five days, had there been no necessity for husbanding our stock. The mutineers must naturally have concluded that we could have no other place of refuge than the Friendly Islands; for it was not likely they should imagine, that, so poorly equipped as we were in every respect, there could have been a possibility of our attempting to return homewards: much less will they suspect that the account of their villany has already reached their native country.

1789. June 14.

When I reflect how providentially our lives were saved at Tofoa, by the Indians delaying their attack, and that, with scarce any thing to support life, we crossed a sea of more than 1200 leagues, without shelter from the inclemency of the weather; when I reflect that in an open boat, with so much stormy weather, we escaped foundering, that not any of us were taken off by disease, that we had the great good fortune to pass the unfriendly natives of other countries without accident, and at last happily to meet with the most friendly and best of people to relieve our distresses; I say, when I reflect on all these wonderful escapes, the remembrance of such great mercies enables me to bear, with resignation and chearfulness, the failure of an expedition, the success of which I had so much at heart, and which was frustrated at a time when I was congratulating myself on the fairest prospect of being able to complete it in a manner that would fully have answered the intention of his Majesty, and the honourable promoters of so benevolent a plan.

With respect to the preservation of our health, during a course of 16 days of heavy and almost continual rain, I would recommend to every one in a similar situation the method we practised, which is to dip their cloaths in the salt-water, and wring them out, as often as they become filled with rain; it was the only resource we had, and I believe was of the greatest service to us, for it felt more like a change of dry cloaths than could well be imagined. We had occasion to do this so often, that at length all our cloaths were wrung to pieces: for, except the few days we passed on the coast of New Holland, we were continually wet either with rain or sea.

Thus, through the assistance of Divine Providence, we surmounted the difficulties and distresses of a most perilous voyage, and arrived safe in an hospitable port, where every necessary and comfort were administered to us with a most liberal hand.

Timor.

1789. July.

As, from the great humanity and attention of the governor, and the gentlemen, at Coupang, we received every kind of assistance, we were not long without evident signs of returning health: therefore, to secure my arrival at Batavia, before the October fleet sailed for Europe, on the first of July, I purchased a small schooner; 34 feet long, for which I gave 1000 rix-dollars, and fitted her for sea, under the name of His Majesty's schooner Resource.

July. 20.

On the 20th of July, I had the misfortune to lose Mr. David Nelson: he died of an inflammatory fever. The loss of this honest man I very much lamented: he had accomplished, with great care and diligence, the object for which he was sent, and was always ready to forward every plan I proposed, for the good of the service we were on. He was equally useful in our voyage hither, in the course of which he gave me great satisfaction, by the patience and fortitude with which he conducted himself.

July. 21.

July 21st. This day I was employed attending the funeral of Mr. Nelson. The corpse was carried by twelve soldiers drest in black, preceded by the minister; next followed myself and second governor; then ten gentlemen of the town and the officers of the ships in the harbour; and after them my own officers and people.

After reading our burial-service, the body was interred behind the chapel, in the burying-ground appropriated to the Europeans of the town. I was sorry I could get no tombstone to place over his remains.

This was the second voyage Mr. Nelson had undertaken to the South Seas, having been sent out by Sir Joseph Banks; to collect plants, seeds, &c. in Captain Cook's last voyage. And now, after surmounting so many difficulties, and in the midst of thankfulness for his deliverance, he was called upon to pay the debt of nature, at a time least expected.

August 20.

August the 20th. After taking an affectionate leave of the hospitable and friendly inhabitants, I embarked, and we sailed from Coupang, exchanging salutes with the fort and shipping as we ran out of the harbour.

1789. August.

I left the governor, Mr. Van Este, at the point of death. To this gentleman our most grateful thanks are due, for the humane and friendly treatment that we have received from him. His ill state of health only prevented him from showing us more particular marks of attention. Unhappily, it is to his memory only that I now pay this tribute. It was a fortunate circumstance for us, that Mr. Wanjon, the next in place to the governor, was equally humane and ready to relieve us. His attention was unremitting, and, when there was a doubt about supplying me with money, on government account, to enable me to purchase a vessel, he chearfully took it upon himself; without which, it was evident, I should have been too late at Batavia to have sailed for Europe with the October fleet. I can only return such services by ever retaining a grateful remembrance of them.

Mr. Max, the town surgeon, likewise behaved to us with the most disinterested humanity: he attended every one with the utmost care; for which I could not prevail on him to receive any payment, or to render me any account, or other answer, than that it was his duty.

Coupang is situated in 10° 12´ S latitude, and 124° 41´ E longitude.

August 29.

On the 29th of August, I passed by the west end of the Island Flores, through a dangerous strait full of islands and rocks; and, having got into the latitude of 8° S, I steered to the west, passing the islands Sumbawa, Lombock, and Bali, towards Java, which I saw on the 6th of September. I continued my course to the west, through the Straits of Madura.

September 10.

Passourwang

On the 10th of September, I anchored off Passourwang, in latitude 7° 36´ S, and 1° 44´ W of Cape Sandana, the N E end or Java.

1789. September

On the 11th I sailed, and on the 13th arrived at Sourabya, latitude 7° 11´ S, 1° 52´ west.

Sourabya. Crissey.

On the 17th of September, sailed from Sourabya, and the same day anchored at Crissey, for about two hours, and from thence I proceeded to Samarang. Latitude of Crissey 7° 9´ S, 1° 55´ west.

Samarang.

Batavia.

On the 22nd of September, anchored at Samarang; latitude 6° 54´ S; 4° 7´ W. And on the 26th I sailed for Batavia, where I arrived on the 1st of October. Latitude 6° 10´ S; 8° 12´ W from the east end of Java.

On the day after my arrival, having gone through some fatigue in adjusting matters to get my people out of the schooner, as she lay in the river, and in an unhealthy situation, I was seized with a violent fever.

On the 7th, I was carried into the country, to the physician-general's house, where, the governor-general informed me, I should be accommodated with every attendance and convenience; and to this only can I attribute my recovery. It was, however, necessary for me to quit Batavia without delay; and the governor, on that account, gave me leave, with two others, to go in a packet that was to sail before the fleet; and assured me, that those who remained should be sent after me by the fleet, which was to sail before the end of the month: that if I remained, which would be highly hazardous, he could not send us all in one ship. My sailing, therefore, was eligible, even if it had not been necessary for my health; and for that reason I embarked in the Vlydt packet, which sailed on the 16th of October.

Cape of Good Hope.

1789. December.

On the 16th of December, I arrived at the Cape of Good Hope where I first observed that my usual health was returning; but for a long time I continued very weak and infirm.

I received the greatest attention and politeness from the governor-general, and all the residents on the coast of Java; and particular marks of friendship and regard from the governor, M. Van de Graaf, at the Cape of Good Hope.

On the 2d of January, 1790, we sailed for Europe, and on the 14th of March, I was landed at Portsmouth by an Isle of Wight boat.

FINIS.

Printed in Great Britain
by Amazon